£6.00

Properties of Matter

The Manchester Physics Series

General Editors
F. MANDL : R. J. ELLISON : D. J. SANDIFORD

Physics Department, Faculty of Science,
University of Manchester

PROPERTIES
OF
MATTER

B. H. Flowers, F.R.S.
Langworthy Professor of Physics,
University of Manchester

E. Mendoza
Professor of Physics,
University College of North Wales, Bangor

John Wiley & Sons Ltd.

CHICHESTER   NEW YORK   BRISBANE   TORONTO

Library of Congress Catalog Card No. 70-118151
ISBN 0 471 26497 0 Cloth bound
ISBN 0 471 26498 9 Paper bound

Reprinted February 1978
Reprinted February 1979
Reprinted September 1980
Reprinted May 1982
Reprinted February 1985
Reprinted May 1989
Reprinted May 1990

Printed and bound in Great Britain by
Courier International Ltd, Tiptree, Essex

Editors' Preface to the Manchester Physics Series

In devising physics syllabuses for undergraduate courses, the staff of Manchester University Physics Department have experienced great difficulty in finding suitable textbooks to recommend to students; many teachers at other universities apparently share this experience. Most books contain much more material than a student has time to assimilate and are so arranged that it is only rarely possible to select sections or chapters to define a self-contained, balanced syllabus. From this situation grew the idea of the Manchester Physics Series.

The books of the Manchester Physics Series correspond to our lecture courses with about fifty per cent additional material. To achieve this we have been very selective in the choice of topics to be included. The emphasis is on the basic physics together with some instructive, stimulating and useful applications. Since the treatment of particular topics varies greatly between different universities, we have tried to organize the material so that it is possible to select courses of different length and difficulty and to emphasize different applications. For this purpose we have encouraged authors to use flow diagrams showing the logical connection of different chapters and to put some topics into starred sections or subsections. These cover more advanced and alternative material, and are not required for the understanding of later parts of each volume.

Since the books of the Manchester Physics Series were planned as an integrated course, the series gives a balanced account of those parts of physics which it treats. The level of sophistication varies: '*Properties of Matter*' is for the first year, '*Solid State Physics*' for the third. The other volumes are intermediate, allowing considerable flexibility in use. '*Electricity and Magnetism*', '*Optics*' and '*Atomic Physics*' start from first year level and progress to material suitable for second or even third year courses. '*Statistical Physics*' is suitable for second or third year. The books have been written in such a way that each volume is self-contained and can be used independently of the others.

Although the series has been written for undergraduates at an English university, it is equally suitable for American university courses beyond the Freshman year. Each author's preface gives detailed information about the prerequisite material for his volume.

In producing a series such as this, a policy decision must be made about units. After the widest possible consultations we decided, jointly with the authors and the publishers, to adopt SI units interpreted liberally, largely following the recommendation of the International Union of Pure and Applied Physics. Electric and magnetic qualities are expressed in SI units. (Other systems are explained in the volume on electricity and magnetism.) We did not outlaw physical units such as the electron-volt. Nor were we pedantic about factors of 10 (is 0.012 kg preferable to 12 g?), about abbreviations (while s or sec may not be equally acceptable to a computer, they should be to a scientist), and about similarly trivial matters.

Preliminary editions of these books have been tried out at Manchester University (and in the case of '*Properties of Matter*' also at Bangor University) and circulated widely to teachers at other universities, so that much feedback has been provided. We are extremely grateful to the many students and colleagues, at Manchester and elsewhere, who through criticisms, suggestions and stimulating discussions helped to improve the presentation and approach of the final version of these books. Our particular thanks go to the authors, for all the work they have done, for the many new ideas they have contributed, and for discussing patiently, and frequently accepting, our many suggestions and requests. We would also like to thank the publishers, John Wiley and Sons, who have been most helpful in every way, including the financing of the preliminary editions.

Physics Department	F. MANDL
Faculty of Science	R. J. ELLISON
Manchester University	D. J. SANDIFORD

Preface

A radical revision of the undergraduate physics syllabus of the University of Manchester was undertaken in the year 1959. This exercise involved the participation of many members of the academic staff. It was eventually decided to base the whole of the syllabus upon two introductory first-year courses, one concentrating on the general properties of wave motions, the other based upon the statistical properties of matter considered as a collection of interacting atoms and molecules. The latter course, consisting of about 34 50-minute lectures, was first given in the 1959/60 academic session under the title 'Properties of Matter'; over the years it has developed into the present book.

Our aim has been to show how the macroscopic quantities describing matter in bulk can be related to each other in terms of the microscopic properties of molecules and their interactions. This of course is the subject matter of statistical thermodynamics. To the purist this subject can only be tackled after a thorough grounding in advanced mechanics, thermodynamics and the quantum theory. But the spirit of inquiry amongst undergraduates, and the incentive to devote their time and energies to these rigorous pursuits can more readily be generated, it seemed to us, if they can first be made aware of what much of physics is about in a more rough and ready fashion. It is perhaps contrary to the present fashion, but we have omitted all quantum considerations from the foundations of this work, confining ourselves to a few passages here and there which are in the nature of 'see the next exciting instalment' when, indeed, quantum theory is necessary rather than merely desirable in order to understand some macroscopic phenomenon. Similarly, we have excluded any discussion of the second law of thermodynamics and its consequences—at the risk, here and there, of doing violence to the distinction between internal energy (which we calculate) and free energy (which we do not). We hope that we have at least identified the points at which the distinction matters. However, we consequently have not always been able to avoid the phrase 'It can be shown that . . .', although we have tried to avoid any implication of it except in peripheral matters. We comfort ourselves by suggesting that physics would be very dull unless there were always some things left outstanding in this way.

More importantly, however, we have been forced to restrict severely the number of kinds of matter we were prepared to discuss. We have excluded all discussion of ionized plasmas, of polymers and of biological materials. Each of these, it seemed to us, requires a book to itself. We have touched, although briefly, on the engineering properties of materials limiting ourselves to a discussion of the strength of real solids—for our concern has been rather to show that these properties can in principle be related to the microscopic properties. Argon, gaseous, liquid and solid, figures ubiquitously. It is perhaps the simplest element from our point of view, the ideal element, about which much experimental information is available to us for our simple-minded analysis. Apart from that, we have mostly confined ourselves to gases, liquids and solids consisting of small molecules, and to simple ionic substances and metals.

This is the way in which much of the study of the properties of matter developed historically. We hope that we have succeeded in bringing back some of the excitement of the original discoveries; certainly we found it exciting to rediscover some of these ourselves.

The course has been given in modified form in Manchester since 1959 and in Bangor since 1965, by others as well as by ourselves. We are indebted to several of our colleagues who, as lecturers or tutors, have contributed much to its gradual development. We are particularly indebted to Dr David Caroline, as well as to the editors of the Manchester Physics Series for their friendly but penetrating criticisms and suggestions. Most of all we are indebted to more than a thousand of our students whose own efforts to understand what we were trying to do has been our main encouragement and incentive. They are not, of course, responsible for the remaining imperfections in our book.

B. H. FLOWERS
ERIC MENDOZA

List of Symbols

A, B	constants
A	area
A_0	activation energy
a, a_0	atomic or molecular diameter
$\mathfrak{a}, \mathfrak{b}$	constants in van der Waals' equation
α	linear expansion coefficient; Madelung constant
β	volume expansion coefficient
γ	ratio of specific heats; surface tension
γ_G	Grüneisen constant
C	specific heat, usually with suffix: C_p, C_v
c	speed of molecule; speed of light
D	diffusion coefficient
d	distance
E	energy
e	charge on electron
ε	depth of interatomic potential well
F	force
$f(\ )$	function
$\mathscr{F}$	faraday
η	viscosity coefficient
J	flux of particles
K	bulk modulus
$°K$	degrees absolute
k	Boltzmann's constant
κ	velocity coefficient of chemical reaction
κ	thermal conductivity
$\mathscr{K}$	kinetic energy
L_0	latent heat at low temperatures
$\mathscr{L}$	Lorentz number
ℓ	length
Λ	wavelength
λ	mean free path
M	molecular weight
m	mass of atom

N	Avogadro's number
n	number, number density
$\mathfrak{n}$	coordination number
$\mathcal{N}$	number per unit area
ν	frequency
ν_E	Einstein frequency
P	pressure
$P[\ ]$	probability function
p_x, p_y, p_z	momentum components
p, q	indices of interatomic potential energy
r	radial distance
ρ	density
s	strain
σ	collision cross-section; conductivity
T	temperature
t	time
τ	characteristic time
U_x	drift velocity
V	large volume
V_0	molar volume
v	small volume
v_x, v_y, v_z	velocity components
$\mathscr{V}$	potential energy
ω	angular velocity
★	sections or subsections marked with a star may be omitted, if the reader so wishes, as they are not required later in the book

Contents

★Starred sections or subsections may be omitted, if the reader so wishes, as they are not required later in the book.

8   THERMAL PROPERTIES OF SOLIDS

9   DEFECTS IN SOLIDS: LIQUIDS AS DISORDERED SOLIDS

CHAPTER

1

The study of the properties of matter

1.1 THE STUDY OF THE PROPERTIES OF MATTER

Throughout the whole of the nineteenth century, one of the open questions of science was whether matter was composed of atoms or not. Nobody had yet been able to perform experiments with single atoms, certainly no-one had ever seen one, and for a long time no-one knew even the order of magnitude of the sizes of atoms—whether their diameters were typically of order 10^{-5} cm or 10^{-50} cm. One method of attack was to try and *correlate* as many different properties of solids, liquids and gases as possible on the basis of simple postulates about the forces which atoms exerted on one another. The earliest attempt at describing these forces was made by Boscovitch in 1745. Sixty years later, a triumph was scored when Laplace, arguing from the fact that the rise of a liquid in a capillary tube was observed to be independent of the thickness of the wall of the tube, deduced that atomic forces must act only over short distances. He was able to deduce theoretically the form of the surface-tension law for liquids, that the force exerted by surface tension should be proportional to the length of a cut in the surface—and this was verified experimentally. Much later, in the 1860's and 70's, the transformation of gas into liquid was demonstrated for many substances when it became technically possible to produce high pressures and low temperatures. The similarities and regularities in behaviour of several substances, predicted on the basis of crude atomic models, added plausibility to those models. Above all, the rough agreement between estimates of the *sizes* of atoms based on widely differing kinds of experiments (about eight completely different methods all gave atomic diameters of the order of 10^{-7}–10^{-8} cm) made the atomic hypothesis fairly secure by 1900.

Thus the subject called 'Properties of Matter' or 'Heat' was at one time an exciting one. Physicists measured surface tensions and latent heats and elasticities and tried to correlate them under an all-embracing atomic theory. But with the discovery of sub-atomic particles and the invention of counting devices which could detect single atoms or ions, the subject lost its urgency. By the early years of the twentieth century, no-one anywhere doubted that matter was atomic in structure and that atoms were of the order of 10^{-8} cm in diameter. Experimenters still measured surface tensions, latent heats and elasticities, but these had now become respectable, if routine, activities in their own right. Books came to be written entitled 'Properties of Matter' which described highly sophisticated apparatus for measuring quantities of this kind, and gave elaborate calculations on the twisting of laminas and the bending of beams, but never mentioned the word 'atom'.

It is the purpose of this book to try and recapture some of the spirit of the old approach. We will in fact *start* from statements about the shapes and sizes of atoms and the forces holding them together, and then show how the properties of solids, liquids and gases can be deduced. It is our purpose to show that, given the potential energy between two atoms of known atomic weight, it is possible to estimate the density of the solid and its specific heat, its thermal expansion and elasticity, the surface tension and latent heat and viscosity of the liquid, the diffusion constant and thermal conductivity and specific heat of the gas and the velocity of sound through it: they are all *related* properties of matter.

1.2   ORDERS OF MAGNITUDE

The estimates we shall make will rarely be exact ones. Since the object of our discussions will be mainly to show that we can identify the forces or mechanisms underlying certain phenomena, it will serve our purpose if we can show that using approximate methods we can get *roughly* the right answer. To improve on rough estimates usually demands a great increase in mathematical complexity, and it would achieve little if we risked obscuring the line of the argument by getting involved in complicated manipulations merely to add a few percent to the accuracy of the result. In many operations, it is by contrast extremely important to know

Hydrogen gas

Fig. 1.1. A diagram drawn by John Dalton, in his epoch-making book *A New System of Chemical Philosophy* published in 1810. The gas is shown as a *regular* arrangement of atoms, which is quite wrong. Not till this idea was supplanted could any real progress be made.

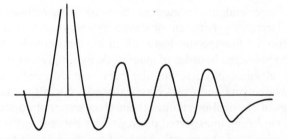

Fig. 1.2. An early idea of the effect of one atom on another—from a book published by Boscovitch in 1745. Compare this diagram with Fig. 3.4. The oscillations in this graph were postulated to account for the structure of a gas as pictured above, but this remains an astonishingly penetrating attempt to explain the properties of matter in fundamental terms.

the *exact* values of certain parameters. Chemical engineers, for example, need to know thermal constants to five-figure accuracy in order to predict whether they can manufacture a certain product economically or not. Similarly, a bridge might collapse if a designer made a mistake by a factor of two in some of his data. But for our present purposes, we will be content if we can estimate that a thermal or chemical change may take place at some temperature *of the order of* a few hundred degrees absolute; and we will regard it as satisfactory if, given the atomic constitution of both, we can predict from first principles that steel is a good deal stronger than butter.

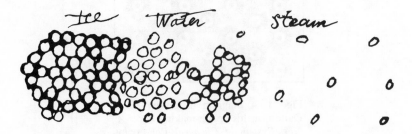

Fig. 1.3. Possibly the very first 'modern' drawing of the molecules in the solid, liquid and gas phases. A 'doodle' from one of Joule's notebooks, done while he was working out the implications of the conservation of energy and realizing that the old static picture of gases was wrong (1847).

This emphasis on the importance of orders of magnitude must not be taken to disparage the crucial role of accuracy in experimental measurements nor to suggest that exact theories need not be pursued. Indeed the existence of new and unexpected phenomena is sometimes shown up when the discrepancy between observation and theory is quite small. Measurements of the specific heats of gases, for example, give results which can be predicted in order of magnitude by simple theories based on the laws of classical physics, but the persistent disagreement between precise measurements and exact classical theories was the first evidence that the laws of classical mechanics themselves were not applicable in all circumstances. Again, unexpected discrepancies between existing theories and measurements of the specific heats of solids at lower temperatures could similarly only be explained by using a quantum approach; measurements on metals showed that electrons deviated sharply from the classical behaviour that had been expected. These important phenomena (which will all be discussed later in this book) would not have been discovered had physicists been content merely to make rough estimates. Nevertheless, if a rough estimate does give a result which agrees in order of magnitude with observation, this can usually be taken to mean that the correct mechanisms have at least been identified. This will be the main theme of this book.

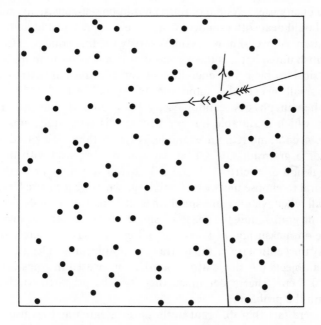

Fig. 1.4. A very early picture, one of the first to be drawn to scale, of a gas—the molecules of air in a volume 10^{-4} cm square by 10^{-8} cm thick—not long after Avogadro's number had been reliably estimated. Kelvin 1883.

1.3   UNITS AND SYSTEMS OF UNITS

When we make a statement about a physical quantity, like 'the mass of a proton is 1.66×10^{-24} g', the datum consists of three parts: the number of order unity, the power of 10 and the unit (1.66, 10^{-24} and the gram, respectively). Many students have a habit of remembering the first figures to high accuracy, but of forgetting the power of 10 or ignoring the units. In our collective memory we can recall students who have insisted that the sun is 3,000 miles from the Earth, that the diameter of an atom is 109,737 cm, that Planck's constant is 6.6×10^{23} unknown units, that gravitation is the cohesive force that holds solids together. Wildly wrong statements like these are sillier than saying that butter is a good material for building bridges. In quoting physical quantities, it is usually more important to get the power of 10 and the units correct—to be sure of the *order of magnitude*—than it is to quote the digits at the beginning. But having got the figures correct, the units must be quoted; data without units are devoid of meaning.

To be of any use, a system of units should be self-consistent and preferably it should deal with numbers which are comprehensible to the human imagination. All well-known systems satisfy the first condition (S.I., c.g.s., even British units) but it is the second that it is difficult to fulfil. In almost any physical problem, one encounters numbers which are extraordinarily large or small by everyday standards; numbers like 10^{23} or 10^{-16} which occur often in physics, cannot easily be visualized. Further, any unit quantity which is suitable for one problem is often quite unsuited for another. A coulomb (C), for example, is a tiny thing compared with the charge on a gram ion (10^5 C) but enormous compared with the charge on a single ion or electron (10^{-19} C). To avoid very large or small factors, it is natural to choose units which are large when we measure large things and small when we measure small ones. It is natural to measure atomic weights in grams, and the mass of single atoms in atomic mass units; to insist on measuring *all* masses in kilograms, say, is merely perverse. Similarly the centimetre or the metre are suitable units of length for many common objects but single atoms are best measured in Ångstrom units: $1 \text{ Å} = 10^{-8}$ cm. The reader must, therefore, be prepared to change his units with the problem.

It is a sad fact that different authors of research articles use different *systems* of units, where not only do the symbols stand for different magnitudes and dimensions, but different numerical constants also appear. In electromagnetic equations, statements of the same equation may or may not contain factors like c, 4π, $\varepsilon_0\mu_0$ which arise from the units, and make them difficult to understand. In fact, few difficulties of this kind appear in this book: but the competent physicist must be prepared to be able to read papers written by authors who may be working in any system, and the student must be facile in all of them.

1.3.1   Energy units

The joule (J) is the common unit of energy. For some problems it is a suitable unit since the gas constant R (defined in section 4.4.2) is 8.31 J per degree and this is not a large number. For measuring molar binding energies or latent heats or heats of reaction, the kilojoule (10^3 J) is more appropriate. For measuring the corresponding energies of a single atom or molecule, however, which are roughly 10^{24} times smaller, the joule is not appropriate. Nor is the common 'small' unit of energy, the erg, because it is only a factor of 10^7 times smaller than the joule and one finds oneself dealing with awkward-sounding amounts of energy such as 10^{-11} or 10^{-14} erg. There does in fact exist an energy unit which is suitable for these purposes. It is the electron volt. This is the amount of energy acquired by an electronic charge when it falls through a potential difference of 1 volt: 1 electron volt (1 eV) $= 1.60 \times 10^{-12}$ erg $= 1.60 \times 10^{-19}$ J. (This relation can be calculated from the knowledge (a) that when 1 coulomb falls through 1 volt, 1 joule of energy is released, and (b) that the charge on an electron (section 2.1.2.) is 1.60×10^{-19} C). In these units molecular binding energies lie between 10 and 0.01 eV, and these numbers are easy to visualize and handle.

2

Atoms, molecules and the states of matter

2.1 ATOMS, IONS AND MOLECULES

It is convenient for us to start with the statement that matter is composed of atoms. Atoms are not the fundamental units of nature because
they themselves can be broken up into a few smaller, and in a sense more
fundamental, units or elementary particles (electrons, neutrons and
protons). But the conditions needed to make atoms disintegrate are
rather extreme and are not normally met with (at any rate if we except
radioactive substances whose nuclei disintegrate but which do not concern
us particularly), so that from the present point of view we need go no further
than to say that matter is built up of atoms.

Under ordinary conditions, matter seems to be continuous. Given a
small piece of any solid, for example, it is possible to cut it up into smaller
fragments and to go on repeating this process; there seems to be no limit
to the fineness of subdivision, other than that set by the instruments
available. But in fact (if we carry out the process by *any* method, under
conditions of temperature and pressure which are not too extreme) there
is a limit when we reach atomic dimensions. The illusion that matter is
continuous is due to the extreme smallness of even the largest atoms, and
to the very large numbers of them which are present even in a microscopic
speck.

We shall be concerned mostly with the forces between atoms, and it is therefore necessary to describe their structure and the formation of ions and molecules, so that the origin of the forces which they exert on one another can be understood.

2.1.1   Atomic number

If a sample of a substance can be shown to consist of atoms all of one kind, that substance is said to be a chemical *element*. Three examples of elements are hydrogen (H), under normal conditions a gas which easily takes part in a number of chemical reactions with other elements; helium (He), a gas which is chemically inert and hardly reacts at all with other elements; and lithium (Li), a highly reactive metal.

It is possible to find similarities and regularities among the physical and chemical properties of elements. Of the many characteristics which it is possible to select and use to arrange the elements in some sort of order, one has been found to have a special significance. When samples of the elements are bombarded with energetic electrons, X-rays are emitted, any element giving a spectrum containing many characteristic wavelengths. Each of them, however, includes a recognizable group of four lines, called the *K*-lines, whose wavelengths vary from element to element. The lines from hydrogen have the longest wavelength, those from helium the next longest, then lithium, and in this way it is possible to arrange the elements in order. On this basis, hydrogen is said to have atomic number 1, helium 2, and so on.

2.1.2   Structure of atoms

An atom consists of a nucleus, consisting of neutrons and protons, which is extremely small (about 10^5 times smaller in diameter than the atom as a whole) but which contains almost all the mass. Around this nucleus is a cloud of electrons. This cloud is easy to visualize when there are many electrons; but it is a fact, made comprehensible by quantum mechanics, that a single electron in an atom also behaves somewhat as if it were spread tenuously throughout a certain volume. Even though the electron is a point charge, it appears to an outside observer as if it were continuously spread out. The term 'electron cloud' is therefore appropriate even to an atom containing only a single electron. Each electron carries a negative electric charge, whose *magnitude* is a fundamental constant:

$$e = 1.602 \times 10^{-19} \text{ C}.$$

It is found that the number of electrons inside any atom of an element is equal to the atomic number of that element. To maintain the electrical neutrality of each atom, the nucleus is positively charged, with a magnitude equal to that of all the electrons in the cloud outside it. Thus a hydrogen atom has one negatively charged electron surrounding a nucleus which has one unit of positive charge. Each helium atom has two electrons outside a nucleus carrying two units of charge, and so on.

Between each electron and the nucleus there exist forces of attraction (Coulomb forces) which bind electrons to the nucleus—that is, they make it difficult for the electrons to escape. However, some electrons may be more tightly bound than others—they require greater energy to separate them from the nucleus—and at the same time the clouds are usually smaller in size. In helium and the other 'inert' or 'rare' gases such as neon and argon which resemble it, all the electrons are tightly bound and no further electrons can be added to the system if it is to remain stable. In lithium and the other alkali metals like sodium and potassium and the 'noble' metals copper, silver and gold, most of the electrons are tightly bound but there is one which is rather loosely bound, so that it does not take much energy to detach it from its atom. In hydrogen also, the single electron is only loosely bound. These elements are said to be monovalent and to contain one valence electron. Other elements have more than one loosely bound electron in each atom; the alkaline-earth metals such as beryllium and magnesium each have two. In another group of elements, the halogens, which include the gases fluorine and chlorine, all the electrons are tightly bound but it is possible for another single electron to enter the existing cloud and become tightly bound too. Other atoms can accept more than one extra electron in this way; oxygen for example can accept two, nitrogen three. Only the loosely bound electrons can enter into combinations in this way. The tightly bound ones remain undisturbed by ordinary chemical changes.

2.1.3 Ions

These are atoms which have lost or gained one or more electrons (while keeping their nuclei unchanged) so that they are no longer electrically neutral. An atom of hydrogen can lose its electron (which is of course negatively charged) so that it has an excess of one unit of positive charge. Its mass is very little different from that of a hydrogen atom since most of the mass resides in the nucleus which is unchanged. A chlorine ion is formed by adding one electron to the atom, oxygen can form two kinds of ions according as one or two electrons are added to the atom. The properties of ions are quite different from those of atoms not only because ions are charged but also because in ions all the electrons are tightly bound. The relative sizes of atoms of mercury, of a free ion and of the ions in metallic mercury are shown in Fig. 2.1.

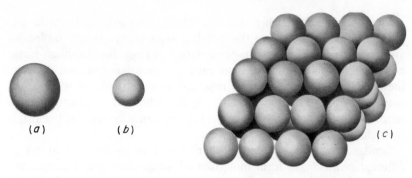

Fig. 2.1. Scale drawings of (a) a mercury atom in the vapour at room temperature (4.4 Å in diameter), and (b) a mercury ion (Hg^{++}) formed from an atom which has lost two loosely bound electrons (2.24 Å in diameter). (c) The ions in the solid or liquid metal are a little bigger than this.

Free ions can be formed in two ways. They can be produced by chemical action, when the initially neutral atom is very close to other atoms which contribute or take away electrons. They can also be produced when neutral atoms are bombarded by beams of particles such as energetic electrons, when the electrons inside the atoms are knocked out by what can be regarded as direct collisions.

Any mass of matter in equilibrium must be electrically neutral, or very nearly so; if any ions are present it is exceedingly probable that oppositely charged ions are to be found not far away.

2.1.4   Molecules

It is possible for the loosely bound electrons to be *shared* in various ways between different atoms, so as to bind those atoms together to form molecules. Some substances do so in the gaseous state but the molecules do not preserve their identity in the solid state; other substances form molecules which exist in all three states, solid, liquid and gas. Hydrogen atoms find it favourable at ordinary temperatures to share their electrons so as to form molecules each containing two atoms. They can be pictured as two nuclei embedded in a cloud of electrons enveloping them both. The molecule forms a separate, stable entity, and at ordinary temperatures hydrogen gas consists almost entirely of such molecules. When cooled to very low temperatures so that the gas liquefies or solidifies, the molecules become quite tightly packed together but they still more or less retain their identity, though a given hydrogen atom may occasionally wander from molecule to molecule. Halogen gases also form diatomic molecules. Helium and the other rare gases are composed of molecules each containing a single atom (so that it is immaterial whether one calls these atoms or molecules).

Metallic elements such as lithium exist as molecules only in the gas phase produced by heating the metal, though many of these molecules are ionized particularly if the temperature is high. Solid lithium metal however consists entirely of ions, each of which has lost one electron; these valence electrons may be regarded as a gas which fills the space between the ions. Other metals have similar structures. The electron gas is mobile inside metals and this confers the electrical conductivity which is their characteristic.

Similar processes of electron sharing can occur between atoms of different elements, to form molecules of compounds. For example, the water molecule is formed from one oxygen atom and two hydrogen atoms. It can be pictured as a rather large electron cloud (formed from those from the hydrogen atoms and the least tightly bound electrons from the oxygen atom) which envelops the oxygen nucleus and its most tightly bound electrons, somewhere deep inside near its centre, with the two hydrogen nuclei a little distance away. Steam at high temperatures consists predominantly of such molecules. In ice, the same molecules can also be distinguished, though the hydrogen atoms spend a proportion of their time wandering from molecule to molecule. Water also contains a proportion of molecules which have split up into ions, consisting of positively charged hydrogen ions and negatively charged hydroxyl ions (oxygen and hydrogen atoms with one extra electron).

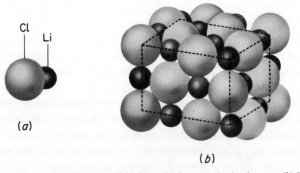

Fig. 2.2. (a) A molecule of lithium chloride which can exist in the gas. (b) In the solid each Li ion is surrounded by six Cl ions—there are no distinguishable molecules. The electron clouds have been drawn with well defined surfaces.

Another simple molecule is that of lithium chloride consisting of a lithium ion bonded to a chlorine ion. These molecules exist only in the gas phase, produced when the substance is vaporized, though even then a number of molecules break up into the constituent atoms or ions. In the liquid and solid, and when dissolved in water, any one ion cannot be said to be definitely associated with any one other ion, so that no molecules exist any more. In the solid, any one lithium ion is surrounded by six chlorine ions, all symmetrically disposed around it, and each chlorine ion is similarly surrounded by six lithium ions. The electrons cannot be accelerated so the substance is an insulator. In dilute solutions, every ion is separately covered by a layer of water molecules. In neither case can it be said that the compound exists in the form of molecules.

Carbon can enter into the composition of a vast number of molecules of many varied shapes, sizes and properties and these constitute organic compounds. They include biological substances whose molecules are often of an astonishing complexity. Such molecules remain almost unchanged in solid, liquid and gas. The benzene molecule, for example, is a flat ring of six carbon atoms each with a hydrogen atom close to it with electron clouds between, and more tenuous clouds above and below the ring. These molecules are present in benzene whether solid, liquid or vapour.

2.1.5 Nuclear and chemical reactions

Atoms and molecules are not indestructible, since the forces holding them together are not infinitely strong. Atoms can be converted into other atoms by nuclear reactions which alter the nuclei, molecules can be changed into other molecules by chemical reactions which affect the loosely-bound electrons and hence alter the groupings of the atoms. However, the forces (or amounts of energy) needed to make nuclei break up or combine are much greater than those required to promote chemical changes. This is made manifest by the temperature which must be attained before reactions of different kinds can be initiated—we shall see later that these temperatures on the absolute scale are a rough measure of the energies involved. Nuclear reactions, such as the combination of hydrogen nuclei to form helium, can only be made to take place at very high temperatures of the order of those found in the interior of stars, but many chemical reactions will proceed at ordinary temperatures, 10^3 or 10^4 times lower on the absolute scale. This topic is discussed again in section 4.4.3.

2.1.6 Atomic masses and Avogadro's number

Atomic and molecular masses can be determined to good accuracy by studying the masses of substances taking part in chemical reactions, and making suitable assumptions as to the nature of the reactions. Nowadays, atomic masses are measured to a high degree of precision by mass spectrometers. The vapour is ionized by bombardment with a beam of electrons and the charged ions are deflected by electric and magnetic fields in such a way that the ratio of charge to mass can be determined. Since the charge must be equal to that of an electron or a multiple of it, the mass can be found.

Observations with mass spectrometers showed that any element, as found naturally, contains atoms of different masses but having the same chemical properties. These are called isotopes of the element. It is the nuclei which differ from one another in mass although they all carry the same charge and the electron clouds round them are identical.

To construct a scale of atomic masses, we take the mass of a given isotope as a standard and compare others with it. The mass of the most abundant isotope of carbon is conventionally taken as 12 atomic mass units, written 12 a.m.u. Natural carbon is a mixture of isotopes with masses close to 12 and 13 a.m.u. and the average is 12.011 a.m.u. The mass of the average hydrogen molecule is 2.016 a.m.u; it is a mixture of atoms of masses near 1 and 2 a.m.u. Oxygen has isotopes of masses close to 16, 17 and 18 a.m.u. Thus the water molecule can have masses anywhere between 18 and 24 a.m.u. the value 18 being overwhelmingly the commonest. For many purposes, however, it is sufficient to quote average masses, rounded off to the nearest whole number. If the mass of a single molecule is M a.m.u. we define one *gram molecule*, often called one mole, of that substance to be M gm. The *molecular weight* of the substance however is usually denoted by M, without any units.

It follows from this definition that the number of molecules contained in one mole of *any* substance is independent of its chemical composition or physical form. This number is called Avogadro's number and is given by

$$N = (6.02257 \pm 0.00009) \times 10^{23}$$

which for most practical purposes can be taken as

$$N = 6 \times 10^{23}.$$

This is the number of hydrogen molecules in 2.016 g of hydrogen, of carbon atoms in 12.011 g of carbon, of water molecules in 18.015 g of water or ice or steam, and so on. The value of the atomic mass unit in grams (the mass of an imaginary molecule of molecular weight equal to 1) is the reciprocal of N:

$$1 \text{ a.m.u.} = (1/N)\,\text{g} = 1.66 \times 10^{-24}\,\text{g}.$$

It is the enormous magnitude of Avogadro's number, which implies that the individual atoms are so extremely small, that gives the illusion that matter is continuous on the ordinary scale.

The gram ionic weight can be defined in a similar way to the gram molecule. The total charge on a gram ion of any substance in which each ion carries a single electronic charge e is Ne units of charge. It is called the faraday and is equal to

$$1\mathscr{F} = 1.60 \times 10^{-19} \times 6.023 \times 10^{23} = 0.965 \times 10^5 \text{ C}.$$

For most practical purposes 1 faraday can be taken as 10^5 C.

So far, we have made several dogmatic statements about the atomic nature of matter, the masses of single atoms and the value of Avogadro's number. Of course, these facts are not self evident and historically they took a long time to prove and were bitterly disputed till surprisingly recent times. In the nineteenth century the proof that the atomic hypothesis was correct hinged on the observation that Avogadro's number could be measured in a large number of completely independent ways and in spite of the crudeness of the measurements, the result was always *roughly* the same. The earliest method involved a study of the rate of diffusion of gases combined with crude estimates of the volumes that the same gases would have occupied if they could have been liquefied or solidified. Other early methods depended on studies of the heat of formation of brass; of the relation between the surface tension and latent heat of liquids; of the colour of the sky and the absorption of starlight in the atmosphere; of the charge on the electron and its relation to electrochemical changes; other methods were also used. All gave values for N which at any rate had the same number of zeros after the first figure and in view of the extraordinary range of phenomena covered, this constituted convincing proof of the atomic hypothesis—though even as late as 1907 it was not universally accepted. Nowadays, it is possible to detect the effect of single atoms and by indirect methods to render them visible, and Avogadro's number is known to high accuracy.

2.2 GASES, LIQUIDS AND SOLIDS

As a useful, though not complete, classification it can be said that matter exists in three states, as gas, liquid or solid. This statement is justified by the fact that there exist many substances which can undergo sharp, easily identifiable, reproducible, and reversible transitions from one state to the other. Water is the classical example: its freezing and melting, boiling and condensation have been contemplated since the time of the ancient Greek scientists. There are obvious contrasts between the properties of ice, water and steam or water vapour which make their description as solid, liquid and gas quite unambiguous. Similarly, most metals are solid, they melt under well defined conditions of temperature and pressure to form liquids and boil at higher temperatures to produce gases.

If all substances possessed such clear demarcations, it would be easy to define the different states of matter. But there are very many substances like glasses or glues which one normally thinks of as being solid but which do not melt at sharply defined temperatures; when heated they gradually become plastic, till they become recognizably liquid. Other solids such as wood or stone are inhomogeneous and it is difficult to describe their structure in detail.

We will therefore not attempt to present definitions of solids, liquids or gases; there would be too many exceptions. Instead we shall describe the principal characteristics of these three states of matter in bulk, and relate them to their structure on the molecular scale. Later we shall describe some of the conditions under which any one state can take on some of the characteristics of the others. In the course of these rather brief summaries, we shall use terms like 'small' or 'large', 'quickly' or 'slowly' which are as yet a little vague. In later chapters, it will be shown how they can be precisely defined.

2.2.1 Compressibility, rigidity, viscosity

Among other properties, it will be useful to compare the compressibility and rigidity or viscosity of the different states of matter. If a given pressure acting on a substance produces a large relative change of volume, that substance is highly compressible. Rigidity is the ability to oppose or withstand forces directed towards changing the shape of a body, while keeping its volume constant; this property refers to a purely static situation. A related quantity is the viscosity, a measure of the resistance to changes of shape taking place at finite speeds. For example, a body moving through a medium has to keep pushing it aside to keep moving; if the forces required to be exerted on the body are large, the medium has a high viscosity.

2.2.2 Properties and structures of gases

Gases have low densities, they are highly compressible over wide ranges of volume, they have no rigidity and low viscosities. The instantaneous structure of a small volume of gas is illustrated in Fig. 2.3. The molecules are usually a large distance apart compared with their diameter and there is no regularity in their arrangement in space. Given the positions of two or three molecules, it is not possible to predict where a further one will be found with any precision—the molecules are distributed at random throughout the whole volume. They are moving randomly with a mean velocity comparable with that of sound, of the order of 10^4 cm/s. Occasionally two or three of them may be found very close to one another so that their electron clouds overlap and they bind together. Such clusters are common at high pressures but they are usually short-lived.

The low density can be readily understood in terms of the comparatively small number of molecules per unit volume, and the high compressibility follows from the fact that the average distance between molecules
can be altered over wide limits. The lack of rigidity can be explained by the
molecules being able to take up any configuration with equal ease. Further,
the molecules can move long distances without encountering one another,
so that there is little resistance to motion of any kind, which is the basis
of the explanation of the low viscosity.

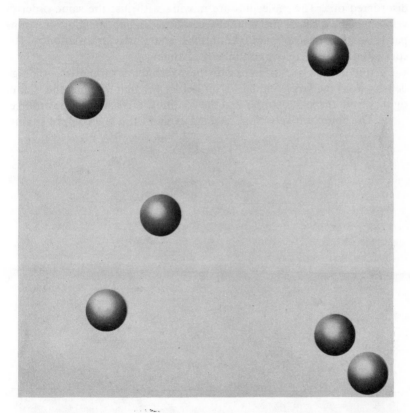

Fig. 2.3. Molecules in a volume of gas at room temperature, a cube of about 20Å at
a pressure of about 20 atmospheres (the molecules are pictured as simple spheres).

2.2.3 Properties and structure of liquids

Liquids have much higher densities than gases—comparing liquids with
common gases under ordinary conditions the factor is of the order of 10^3.
Their compressibility is low. They have no rigidity but their viscosity is of

the order of 10^2 times greater that that of ordinary gases. It is difficult to give a detailed picture of the structure of a liquid; an attempt has been made in Fig. 2.4. The molecules are packed quite closely together so that each one is bonded to a number of neighbours; in the illustration, that number is between 4 and 5. Given the position of one molecule, it is now possible to state how many molecules should be found in contact with it, which is a good deal more information than it is possible to give about the arrangement of the molecules in a gas. But still the pattern as a whole is a disordered one. The molecules are moving with just the same order of velocity as in a gas at the same temperature, though the motion is now partly in the form of rapid vibrations and partly translational. The configuration is therefore continually changing.

This picture can be correlated with the macroscopic properties—the high density from the large number of molecules per unit volume, the lack of rigidity from the lack of order and the continual alteration of the arrangement. The comparatively close packing explains the low compressibility.

Fig. 2.4. Molecules in a volume of liquid about 20Å × 20Å × 3Å.

The fairly high viscosity arises from the fact that the molecules have to wriggle past one another in this irregular but closely packed arrangement, rather like people moving past one another in a dense crowd, where slow relative movements are easy but rapid ones are difficult.

2.2.4   Properties and structure of solids

Solids have practically the same densities and compressibilities as liquids. In addition they are rigid; under the action of small forces they do not easily change their shape.

An important property of those solids which have a well-defined melting-point is that if they are formed very slowly from the liquid state they are crystalline— that is, they form shapes bounded by plane faces with

Fig. 2.5. Molecules in a volume of solid about 20Å × 20Å × 3Å. They are close-packed, and the arrangement is highly regular. There is however a fault (dislocation) in the arrangement towards the left, which can be seen by holding the page horizontal and looking upwards along the rows. (Dislocations are discussed in detail in section 9.3.1).

characteristic angles between them. Sometimes however the crystalline form is not obvious, especially if the solid is not produced under suitably controlled conditions; the crystals may be too small to be seen so that the solid as a whole does not show the expected facets. Substances which do not melt sharply but show a gradual transition to the liquid when heated are said to be amorphous and show no trace of regularity of external shape.

In crystalline solids, the molecules are arranged in regular three dimensional patterns or lattices, of which Fig. 2.5. is an example. If the crystal has been carefully prepared, the regular arrangement persists over distances of several thousand molecules in any direction before there is an irregularity, but if it has been subjected to strains or distortions the regular arrangement may be perfect and uninterrupted only over much shorter average distances. In metals the ions are closely packed together, so that the distance between the centre of an ion and that of one of its nearest neighbours is equal to the diameter of one ion, or something close to it. In other crystals, the packing together of the molecules may be relatively open, but even in a light solid such as ice the distance between the centres of any molecule and its near neighbours is only twice the diameter of a molecule. In solids, the molecules are again moving with the same order of magnitude of velocity as in gases or liquids, but the motion is confined to vibrations about their mean positions.

Amorphous solids can be described as liquids of extremely high viscosity, which over long periods of time have been 'frozen' into one particular configuration. Figure 2.4 could be taken to illustrate the arrangement of molecules in such a solid, with the great difference that the configuration hardly alters with time, the atoms hardly ever changing their relative positions, although they continue to vibrate.

In terms of these structures, it is not surprising that solids have compressibilities and densities like those of liquids. The high rigidities can also be understood, for the molecules can only move with respect to one another with difficulty. In crystalline solids, changes of shape can only take place through molecules slipping into holes which exist at irregularities in the lattice and this is not easy. In amorphous solids, there are plenty of holes for the molecules to slip into in order to initiate a change of shape, but the bonds between neighbours cannot be easily broken by external forces acting on the solid as a whole. All solids therefore resist the action of external forces and this is just what we mean when we say they are rigid.

2.2.5 Characteristic times

There are some substances which are undeniably solids, yet which over a long period of time alter their shape under the action of only small forces. Glaciers, for example, although made of ice which is undoubtedly solid and crystalline, are found to be flowing slowly downhill if they are observed over a period of years. Lead is another well-known example, a metal of high density but comparative softness. Sheets of lead, sometimes used for covering the roofs of large buildings, will alter their shape over a period of decades, slowly creeping downwards under the action of their own weight. Under large pressures, a number of metals flow quite quickly and their ability to fill small cracks makes them suitable for use as washers or gaskets in situations where more fluid sealing compounds cannot be used. Indium and pure gold, as well as lead and other soft metals, can be used in this way. Thus, if we observe their behaviour for short times under the action of small forces, we class these metals as solids; but if we study them under high pressures, or for times of the order of decades (10^8–10^9 seconds) if only small forces act on them, we say that they behave to some extent like liquids.

By contrast, liquids and gases show resistance to bulk motion whose speed is comparable with the speed with which the molecules are moving. Water is well known to feel like a solid if one dives on to it instead of through it. If a gas is made to move with a speed comparable to the velocity of sound it can sustain very sharp changes of density; instead of being uniform, it is divided into distinct regions of different temperatures and densities. The boundaries between them are called shock waves. Thus liquids or gases subjected to large forces for short periods of time exhibit some of the characteristics of solids.

Descriptions of the properties of solids, liquids and gases should therefore include estimates of the times over which it is necessary to extend the observations in order to decide whether a substance has rigidity or not, whether local variations of density can be sustained or not. Under ordinary conditions of temperature and pressure, lead is a solid if we are concerned with events taking place in times which are less than, say 10^7 seconds, but for experiments lasting more than, say 10^9 seconds, lead is a liquid. Similar but much shorter characteristic times can be defined for substances which are gases or liquids. In a similar way, when we say that a substance softens or melts over a certain range of temperature, we mean that below that range the characteristic time for flow under small pressures is very long, inside the range it decreases with rising temperature, and at higher temperatures it is small.

CHAPTER

3

Interatomic potential energies

3.1 MOLECULAR DIMENSIONS

A rough calculation of molecular dimensions can be made if the molar volume of a solid or liquid is known. This is given by

$$V_0 = M/\rho$$

where M is the gram molecular weight and ρ the density in g/cm^3.*

For water, $M = 18$ g and ρ is about 1 g/cm^3 for both liquid and solid. V_0 is therefore about 18 cm^3. Now this is the volume occupied by $N = 6 \times 10^{23}$ molecules. Thus the average volume occupied by a single molecule is 3×10^{-23} cm^3. If we regard this volume as a cube, its side must be about 3×10^{-8} cm $= 3$ Å. If it is a sphere or any other simple shape its linear dimensions will not differ much from this. The distance between molecules in water or ice must therefore be about 3 Å. It has already been mentioned (section 2.2.4) that the lattice in ice and the packing in water are relatively open, so that this figure is greater than the diameter of a water molecule.

In metals, the packing is usually very close, so that the diameter of an ion cannot differ much from $(V_0/N)^{1/3}$. For potassium, a light metal with large ions, $M = 39$ g, $\rho = 0.86$ g/cm^3, so that $V_0 = 45.4$ cm^3 and the

---

* Many students firmly believe that the molar volume of all substances is 22.4 litres. This is indeed roughly true for *gases under 'standard' conditions* of temperature and pressure, 0 °C and 1 atmosphere. It is NOT true for solids or liquids.

diameter of an ion is about 4.2 Å. For gold, one of the densest metals, $M = 197$ g, $\rho = 19.3$ g/cm^3; $V_0 = 10.2$ cm^3 and the ionic diameter = 2.6 Å. Indeed the diameters of all monatomic ions or molecules and the mean diameters of the smaller polyatomic molecules are all between 1.5 and about 5 Å. From the fact that the densities of common gases are of the order of 10^{-3} g/cm^3 (of the order of grams per litre) it follows that the mean distance between molecules is of the order of a few times 10 Å, which is much greater than the diameter of a molecule. This agrees with the situation pictured in Fig. 2.3.

3.2   INTERACTIONS BETWEEN ELECTRICALLY NEUTRAL ATOMS AND MOLECULES

In the previous chapter we described the different spatial arrangements which the molecules or ions could take up in solids, liquids and gases. These can be related in terms of the forces which the molecules or ions exert on one another as a function of their distance apart. In this section we will concentrate on electrically neutral atoms and molecules: metals will be dealt with in section 3.7, ions in section 3.8.

We have to reconcile two apparently contradictory statements. First: liquids and solids are highly incompressible. From this it follows that when two molecules are squashed together so that they approach one another closely, they repel one another. Second: solids and liquids cohere—that is, their molecules tend to pull themselves close to one another. It takes force to stretch a solid, therefore molecules attract one another. We will denote the repulsive forces by F_R and the attractive forces by F_W.

F_R must be dominant when the distance between molecules is about 1–2 Å or less. F_W must be dominant when the separation is about 2–3 Å or greater. At some intermediate distance, say roughly 2 Å, the two forces must be equal and opposite—repulsion balances attraction. (Obviously these figures must not be taken too literally. They are quoted as crudely typical.) We seek an algebraic representation for this.

First we must have a sign convention for the direction of the force exerted by a molecule. Imagine the origin of coordinates to be taken at the centre of the molecule and a line drawn outwards towards another molecule on which it exerts a force. Call this the r-axis. A force which acts in the direction of r increasing—a force of repulsion, tending to separate the molecules—is reckoned positive (Fig. 3.1). A force of attraction is, on the same convention, reckoned negative in sign. We will use this convention consistently.

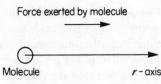

Fig. 3.1. Convention for directions of axes and forces. A repulsive force exerted by a molecule, in the direction of r increasing, is positive.

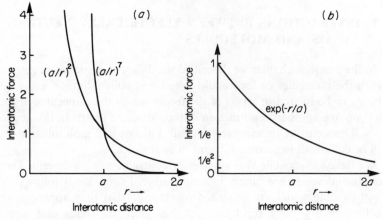

Fig. 3.2. (a) Short range and long range forces; both are repulsive. (b) Exponential fall-off. Again, a repulsion is shown. (Plotted in arbitrary units).

All intermolecular forces, attractions and repulsions, become smaller as the separation increases. If we choose to represent this variation by a simple power law, this law might be

$$\text{Force} = \pm(\text{const.})\left(\frac{a}{r}\right)^{n},$$

choosing the $+$ sign for a repulsion and the $-$ sign for an attraction. a is some standard length and the index n is positive. Now if n is large, r^{-n} becomes rapidly smaller when r is increased and rapidly bigger when the distance is decreased. If n is small, however, then the force falls off comparatively slowly at large distances though it also increases comparatively slowly at small distances. To take a specific example, Fig. 3.2(a), consider two forces proportional to $(a/r)^7$ and $(a/r)^2$ and called F_1 and F_2 respectively. The index 7 is for present purposes a large number and the 2 is a small number. At $r = a$, both forces are numerically equal. However at a small separation $r = a/10$, F_1 is 10^5 times bigger than F_2; but at a large distance $r = 10a$, F_2 is 10^5 times bigger than F_1. A force like F_1, dominant at short distances but negligible at large, is called a *short range* force. F_2 is a *long range* force. Besides simple power laws like $(a/r)^n$ other forces are found in nature which vary like $\exp(-r/a)$ where a is some characteristic distance. Every time r is increased by a, this kind of force decreases by a factor $e = 2.728$, so that it falls to $1/20$ of its value for every $3a$ (Fig. 3.2(b)). Thus exponentially varying forces are certainly short range.

In this terminology, the repulsive forces between atoms are short range, the attractive contributions are rather longer range. The best expression for the total force has been shown theoretically to be of the type

$$F = F_R + F_W$$
$$= A \cdot e^{-r/a} - B(a/r)^7,$$

where the variable r is the distance between the centres of the atoms. a is some measure of the 'diameter' of an atom—not, of course, that it has a sharply defined surface like a billiard ball.

The repulsive force is caused by the overlapping of the two electron clouds—this gives the exponential variation. The attractive force is called the van der Waals force. It arises from the distortion of the electron cloud of one molecule by the presence of the other. It exists even though the atoms are electrically neutral.*

For many purposes it is an advantage to have an expression for the interatomic force which fits the true curve adequately but which has a simple analytical form. In fact, the exponential term can be replaced by one of the type $(a/r)^n$ where n is 10 or 13 or some number like that, so that

$$F = A\left(\frac{a}{r}\right)^n - B\left(\frac{a}{r}\right)^7.$$

This is good enough for our purposes and we will adopt it.

3.2.1   Potential energy

Rather than deal with forces, it is more convenient to deal with the potential energy of two molecules with respect to each other. This is a

---

* Many students firmly believe that the only force of attraction which can exist between electrically neutral atoms is the Newtonian gravitational attraction due to the masses. They assume that the attractive forces between atoms are gravitational. This assumption was made in the early nineteenth century by Dalton and other pioneers. Apart from the fact that the index n is wrong (2 for gravity, 7 for interatomic forces), the magnitude of the binding energy (see section 3.3) is a factor of about 10^{30} times too small:

$$1,000,000,000,000,000,000,000,000,000,000$$

times too small. Indeed, gravitational forces can safely be forgotten in all problems which do not involve the Earth or bodies of comparable mass.

scalar quantity and therefore simpler to discuss than forces, which are vectors.

With the sign convention of the last section—F positive if it acts in the direction of r increasing—the potential energy can be defined by the equation*

$$F(r) = -\frac{d}{dr}\mathscr{V}(r), \tag{3.1}$$

where $F(r)$ means 'the force which depends on r' and $\mathscr{V}(r)$ means 'the potential energy which depends on r'. The force is the gradient of the potential energy. Alternatively,

$$\mathscr{V}(r) = \mathscr{V}(r_0) - \int_{r_0}^{r} F(r) \cdot dr, \tag{3.2a}$$

where r_0 is a standard point and $\mathscr{V}(r_0)$ is the potential energy there. It is usually most convenient to take this constant $\mathscr{V}(r_0)$ as zero; if this is done then we have

$$\mathscr{V}(r) = -\int_{r_0}^{r} F(r) \cdot dr. \tag{3.2b}$$

We are at liberty to choose the standard point r_0 where we please. For many problems it is convenient to take $r_0 = \infty$. Then the potential energy at r is given by

$$\mathscr{V}(r) = -\int_{\infty}^{r} F(r) \cdot dr. \tag{3.2c}$$

In some problems however it is convenient to define $\mathscr{V}$ to be zero at the origin, that is to take $r_0 = 0$. The latitude in the absolute value of the potential energy causes no difficulty because we are usually concerned with measuring only *changes* of potential energy.

As an example, consider the potential energy of two atoms derived from the force of repulsion

$$F = \frac{A}{r^n},$$

where $n > 1$. Then

$$\mathscr{V} = -A\int_{r_0}^{r}\frac{dr}{r^n} = \frac{A}{n-1}\left[\frac{1}{r^{n-1}}\right]_{r_0}^{r}.$$

---

* Really the equations should be written in vector notation

$$\mathbf{F}(\mathbf{r}) = -\mathbf{grad}\,\mathscr{V}(\mathbf{r}); \qquad \mathscr{V}(\mathbf{r}) = \int \mathbf{F} \cdot d\mathbf{r}$$

but as we are dealing with central forces the simpler formulation is adequate.

It is obviously convenient to take $r_0 = \infty$ and to write

$$\mathscr{V}(r) = \frac{A}{n-1}\frac{1}{r^{n-1}}.$$

A graph of this function resembles Fig. 3.3(a). The physical situation being described is that initially two atoms are an infinite distance apart and they are pushed together infinitely slowly so that finally they are separated by r. The potential energy is then increased by this amount. Whether one says that the energy of the second atom is increased with respect to the first, or vice versa, is irrelevant; in fact it is better to talk about the potential energy of the whole system. The energy in fact resides in the field of force between the atoms.

The position of stable static equilibrium of a system occurs when the forces are zero, and this is where the potential energy is a minimum. Left to itself, a system will always move so as to reduce its potential energy. It follows straightforwardly from the definitions that a repulsive force which decreases with distance always has a $\mathscr{V}(r)$ curve of the type shown in Fig. 3.3(a)—or Fig. 3.3(b) which differs from it merely in the addition of an arbitrary constant—so that when the system moves to reduce $\mathscr{V}$, it does so by increasing the separation r. Conversely, attractive forces have rising curves like Fig. 3.3(c).

A system having a potential energy which is a function of displacement like Fig. 3.3(d) will tend to move into the position of minimum energy, where

$$\frac{d\mathscr{V}}{dr} = 0.$$

This is the position marked r_0 in the diagram. The system is then said to be in a potential 'well'. It is called a well because it looks like a hole in the ground. Work must be done, energy must be supplied, to get the system out of the well. The amount of energy needed to take the system far to the left would be ε_1; to move it far to the right, ε_2.

3.2.2   Interatomic potential energy

The interactions between atoms can be represented by a graph of their potential energy as a function of the distance between their centres. Following the argument of section 3.2, it can be expected to be of the form

$$\mathscr{V}(r) = \frac{\lambda}{r^p} - \frac{\mu}{r^q},$$

where p is approximately 9 or 12, q is a smaller number and λ and μ are constants.

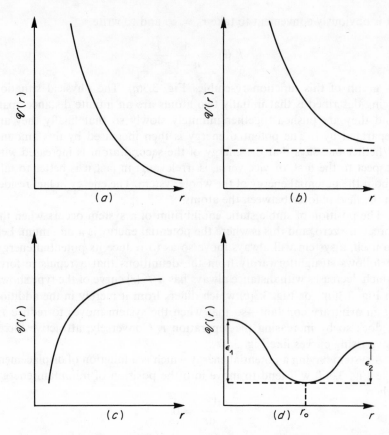

Fig. 3.3. (a) $\mathscr{V}(r)$ for a repulsive force. (b) The same situation with a shifted arbitrary zero for $\mathscr{V}$. (c) $\mathscr{V}(r)$ for an attractive force. (d) A potential well.

A very convenient form of this type of equation is

$$\mathscr{V}(r) = \left(\frac{pq}{p-q}\right)\varepsilon\left\{\frac{1}{p}\left(\frac{a_0}{r}\right)^p - \frac{1}{q}\left(\frac{a_0}{r}\right)^q\right\}. \qquad (3.3)$$

This looks a good deal more complicated than the equation just above it but in essence it is the same, namely the sum of a $1/r^p$ term and a $-1/r^q$ term. A graph of this kind of function (for the special case of $p = 12$, $q = 6$) is shown in Fig. 3.4 and another example ($p = 11$, $q = 1$) is shown in Fig. 3.15(a).

The reader should check that Eq. (3.3) has the following properties. The potential energy at $r = \infty$, when the molecules are infinitely far apart, is zero. By putting $d\mathscr{V}/dr = 0$, it can be readily verified that the minimum value of the energy is $-\varepsilon$, so that we can say that the depth of the well is ε. This minimum occurs at $r = a_0$. This is the position of static equilibrium. In order to pull the atoms apart to infinity, the attractive forces would have to be overcome and this would require the expenditure of an amount of energy ε.

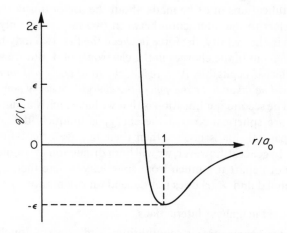

Fig. 3.4. Interatomic potential energy, Eq. (3.3), plotted for the important case of $p = 12$, $q = 6$ (the Lennard–Jones 6–12 potential). The potential energy has been plotted in units of ε, the separation between centres in units of a_0. The curve crosses the axis at $r = a$ (Eq. 3.5).

One case which is specially useful for simple molecules is the *Lennard–Jones 6–12 potential* when $p = 12$, $q = 6$: it reduces to

$$\mathscr{V}(r) = \varepsilon\left[\left(\frac{a_0}{r}\right)^{12} - 2\left(\frac{a_0}{r}\right)^6\right]. \tag{3.4}$$

This is the function plotted in Fig. 3.4. There is a more symmetrical form of this equation, which introduces another parameter a:

$$\mathscr{V}(r) = 4\varepsilon\left[\left(\frac{a}{r}\right)^{12} - \left(\frac{a}{r}\right)^6\right]. \tag{3.5}$$

This is exactly the same as Eq. (3.4), if we put $a_0 = \sqrt[6]{2}\,.\,a = 1.12a$. When the separation r between the centres is equal to a, the potential energy is zero; thereafter, if the two atoms are squashed together a little more the potential energy rises steeply; in other words, the force of repulsion increases greatly. If we regarded the atom as a kind of ball with a hard surface, we would identify the *diameter* of the ball with the separation between centres at which the repulsion rises steeply. Fig. 3.4 would then be interpreted to mean that the diameter of one atom is a and that the position of static equilibrium occurs when the separation is $a_0 = 1.12a$, so that atoms are 'nearly touching'. Since a_0 and a are nearly equal to one another, both are good measures of the *diameter* of one atom.

Two qualifications must be made about the use of this curve. The first is that it refers to the interaction between two molecules only. If a third molecule is in the vicinity, the force between the first two may be modified by the movement of the charges in the electron cloud. But we will assume that this effect is negligible. As a result, *the total energy of an assembly of molecules will be taken to be the sum of the energies of every pair as given by this curve.* The second qualification is that we have tacitly assumed that the molecules are spherical so that the energy is uniquely defined by their distance apart. If this is not so, then their relative orientations may be important. In general, however, we will limit discussion to simple molecules which do not depart too much from sphericity, where their separation r is easily defined and $\mathscr{V}(r)$ does not depend on orientation.

3.2.3   Nearest neighbour interactions

When two atoms are in equilibrium and 'nearly touching', their potential energy is $-\varepsilon$ (Fig. 3.4). If their separation is approximately doubled, their potential energy decreases to about $-\varepsilon/30$, that is, by a large factor. Whereas two atoms which are *nearest neighbours* are bound together by an amount of energy ε ('bound' in the sense that energy is required to separate them), two atoms which are *next* nearest neighbours are only very loosely bound to each other (Fig. 3.5(a)). This is another way of saying that the van der Waals forces are short range forces—though of course the repulsive forces are of even shorter range.

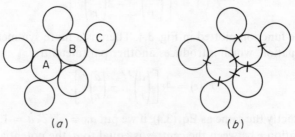

(a)   (b)

Fig. 3.5. (a) Nearest and next-nearest neighbours. The potential energy of A and B is $-\varepsilon$; that of AC is about 30 times smaller. (b) An aid to counting nearest-neighbour interactions. Energy ε is needed to 'cut' each pair apart.

Provided the picture is not taken literally, it can be considered as if each atom were held to its nearest neighbours by some kind of bond which requires an amount of energy ε to cut it (Fig. 3.5(b)). Bonds between atoms which are further apart however are so weak that we can forget them. *The purpose of this picture is to indicate the amount of energy that must be supplied in order to break up the structure. It must NOT be thought to imply that the electrons are more concentrated in certain regions.*

3.2.4   Potential energy dominant at low temperatures

In the rest of this chapter we will show that it is possible to relate some of the macroscopic or large scale properties of solids and liquids to the potential energy between the atoms or molecules.

We will make an important assumption, namely that the kinetic energy of the atoms or molecules is small compared with their potential energy. This is equivalent to saying that we will assume that the temperature is low. At this stage, before we have described in detail what we mean by temperature, it is not possible to say precisely what is meant by a 'low' temperature; in fact, a temperature which is low enough for our approximation to hold for one substance may not be low enough for another. In quoting any data however, we will always take the precaution of referring to temperatures which are low enough for the substance concerned. The effect of this procedure will be to simplify our calculations. In order to estimate the energy of an assembly of a large number of atoms or molecules in a mass of liquid or solid at low temperatures we need only take into account the *potential* energy due to their interactions and we can neglect their kinetic energy.

The result of the discussion of the last section, where we saw that the potential energy of such an assembly is dominated by the potential energy between nearest neighbours simplifies our calculations even further.

3.3   BINDING ENERGY AND LATENT HEAT

The energy required to change one mole of solid or liquid into gas at low pressure is called the binding energy. It is closely allied to quantities which can easily be measured experimentally, the latent heats of *evaporation* (liquid to gas) or *sublimation* (solid to gas).

For precise calculations, there are difficulties when these quantities are compared, however. When we calculate binding energies, we usually take the pressure of the gas to be zero so that the separation between atoms is infinite. Experimentally, we usually take the pressure to be the vapour pressure at the temperature concerned. (For example, one might measure the latent heat to evaporate water at 100 °C to produce steam at 1 atmosphere.) The difference has to be allowed for, although it is small at low temperatures. A more serious source of error is that latent heats of evaporation of liquids are functions of temperature—they *decrease* with increasing temperature and become zero at a high temperature above which the liquid cannot exist. (See for example Fig. 3.13(*b*) for the latent heat of liquid argon.) It is therefore meaningless to quote '*the* latent heat of vaporization' of water or any other liquid as if it were a constant. However, we are restricting our discussion to low temperatures, and in that region latent heats tend towards limiting values: it is these which we consider.

For rough estimates, we note that latent heats of *melting*, to convert solid to liquid, are small compared with latent heats of evaporation to convert liquid to vapour. For example, to convert ice to water at 0 °C requires about 6×10^{10} erg/mol (that is, 340 J/g), to convert water to steam at a comparable temperature requires 45×10^{10} erg/mol (2,500 J/g) —almost eight times as large. So we can approximate even further and, if no better data are available or only rough estimates are needed, we can say that the binding energy is not very different from the latent heat of evaporation.

3.3.1 Estimation of ε from latent heat data

On our approximation, the binding energy at low temperatures is equal to

$$L_0 = \varepsilon \times (\text{number of pairs of nearest neighbours}).$$

It is useful to define the *coordination number* n, the number of nearest neighbours which surround a given atom or molecule. It can never exceed 12. For close-packed solids it can reach 12; for more open arrangements it is smaller, 6 or 10. In dense liquids, the coordination number is about 10. (In the *two-dimensional* pictures, Figs. 2.4 and 2.5, n is about 4 or 5 and about 6 respectively.)

For coordination number n, an assembly of N atoms has $\frac{1}{2}nN$ *pairs* of nearest neighbours; the factor $\frac{1}{2}$ arises from the fact that each bond pictured in Fig. 3.5(*b*) links two atoms but it must only be counted once. Thus:

$$L_0 = \tfrac{1}{2}nN\varepsilon. \tag{3.6}$$

If we know L_0, this allows us to estimate ε. At the same time, the smallness of the latent heat of *melting* can be understood, since the change of coordination number between solid and liquid is quite small (12 to 10 say), whereas between liquid and gas it is large (10 to zero). We can get consistent results for ε by using the latent heat of evaporation together with $n = 10$.*

---

* Strictly our calculations refer to single atoms but they can be applied to molecules which do not depart too far from spherical shape. These include diatomic molecules like N_2 or molecules like CCl_4 which are roughly tetrahedral. Long chain molecules are ruled out.

For liquid nitrogen, consisting of diatomic molecules N_2, the latent heat at low temperatures is about 210 J/g, the molecular weight is 28, so the molar latent heat of evaporation is 6×10^{10} erg/mol. Then ε is 2×10^{-14} erg ~ 0.01 eV. This is the energy needed to separate two nitrogen molecules from one another. For carbon tetrachloride, CCl_4, the latent heat is about 210 J/g, the molecular weight is 153 so that ε is about 10^{-13} erg or 0.05 eV.

These are typical values. Most molecules have ε of the order of 0.01 to 0.1 eV. In Fig. 3.4, each division of the vertical (energy) axis is therefore of this order of magnitude; each horizontal division represents a distance of a few Ångstrom units. Compared with ionization energies, the energies required to remove an electron from the cloud surrounding typical atoms to convert them into ions, these are small amounts of energy. Ionization energies are commonly of the order of 1 to 10 eV, which is 100 to 1,000 times as big.

3.4   SURFACE ENERGY

The surface energy of a solid or liquid is the amount of energy that is needed to create 1 cm^2 of new surface. The process can be pictured as follows. Imagine a column of solid or liquid of 1 cm^2 cross-section to be broken apart by some means, Fig. 3.6. Energy must be used in order to overcome the interactions between molecules on either side of the break. We will calculate this energy. Let there be $\mathcal{N}$ molecules per cm^2 of cross section; if the diameter of one molecule is a_0 cm, then $\mathcal{N}$ is something like $1/a_0^2$ per cm^2.

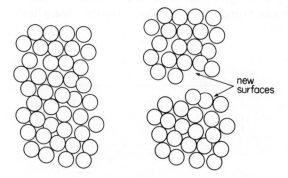

new surfaces

Fig. 3.6. Creating new surfaces by cutting a solid or liquid column in two.

After the break, each molecule in the surface is no longer surrounded by the full number n of nearest neighbours. Instead it has, on the average, only $\frac{1}{2}$n neighbours, in the one hemisphere; therefore $\frac{1}{2}$n$\mathcal{N}$ nearest neighbour interactions must be broken. This requires an amount of energy $\frac{1}{2}$n$\mathcal{N}\varepsilon$: but it produces 2 cm^2 of new surface, 1 cm^2 each for the top and bottom halves of the column. The surface energy is therefore $\frac{1}{4}$n$\mathcal{N}\varepsilon$ erg/cm^2.

3.4.1 Surface tension

Surface tension, a quantity which is easily measured experimentally and is allied to surface energy, is usually defined as the force exerted on a cut 1 cm long in the surface of a solid or liquid, a force which tends to close the cut. It will be denoted by γ dyn/cm. Imagine a thin film of liquid, with upper and lower surfaces like a soap film, to be stretched across a wire frame, Fig. 3.7. One side of the frame is moveable. It is assumed that the film is many molecules thick. The force on the slider is $2\gamma l$ dyn, where l is its length. The factor 2 appears because there are two surfaces. If the slider is moved back a distance d, the work done is $2\gamma l d$ erg. We imagine this process to be done so slowly that heat can flow into the film so that any tendency to cool is counteracted and the temperature and γ remain constant. Since the total area of surface is now increased by $2ld$, the amount of energy supplied is γ erg per unit area. The surface tension is therefore clearly related to the surface energy. They are not identical however, because of the heat energy flowing in during the process to keep the temperature constant. This is the same kind of difference as that between a binding energy and a latent heat. Again, however, if measurements are extrapolated to low temperatures there is little difference between the two, although for the roughest estimates it is not necessary to make even this correction. ((Fig. 3.13(d) shows the variation of surface tension with temperature for liquid argon and this is typical.) We will therefore write

$$\gamma = \tfrac{1}{4}\text{n}\mathcal{N}\varepsilon \tag{3.7}$$

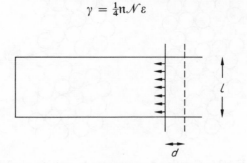

Fig. 3.7. Stretching a liquid film to create new surface.

3.4.2   Estimation of ε from surface tension data

It is clear that, with the considerable oversimplifications in our model of a liquid, we can relate the surface tension to the latent heat. Although the phenomena that we usually associate with these quantities are quite different, both are simply measures of the depth ε of the potential well and of the sizes of molecules. In 1870 Kelvin used a similar analysis to estimate the size of water molecules from the known molar volume V_0, the heat of vaporization and the surface tension, effectively writing our equations in the form $a_0 = 2\gamma V_0/L_0$. This was one of the first methods for estimating Avogadro's number.

Here we will use surface tensions to work out ε for the same liquids as in section 3.3.1 and show that the results are comparable with the values deduced from latent heats. At its normal boiling point (77°K), liquid nitrogen has a density of 0.81 g/cm^3 and its surface tension is 8.7 dyn/cm. Its molecular weight is 28. The molar volume is therefore 35 cm^3, and following the argument of section 3.1.2 the diameter of the molecule is 3.9 Å; hence $\mathcal{N}$ is 6.7×10^{14} per cm². If we take $\mathfrak{n} = 10$, ε must be 0.5×10^{-14} erg or 0.003 eV. This must be compared with 2×10^{-14} erg (0.01 eV) which we deduced from latent heats. For carbon tetrachloride at room temperature the data are 1.6 g/cm^3, 26 dyn/cm and molecular weight 153. The molar volume is 96 cm^3, the diameter of a molecule 5.4 Å, $\mathcal{N}$ is 3.5×10^{14} per cm² and if $\mathfrak{n} = 10$, ε is 3×10^{-14} erg or 0.02 eV. This must be compared with 10^{-13} erg (0.05 eV) from the latent heat data. Both liquids therefore give figures which are consistent within a factor 4, and this must be considered good agreement in view of the crude handling of the data.

★ 3.4.3   The rise of liquids in capillary tubes

One of the commonest methods of measuring the surface tension of a liquid is to measure its rise in a capillary tube of known radius. In this section we will discuss, in terms of interatomic potential energies, why many common liquids rise in a glass tube but mercury falls.

When a glass tube is exposed to an atmosphere containing the vapour of a liquid, its surface is bombarded with molecules and some of these stick to the glass. The process is called adsorption. The whole surface quickly becomes covered with a layer one or two molecules thick. The molecules next to the glass may be attached very firmly: this can be deduced from the amount of energy (the heat of adsorption) which is observed to be given out when the surface is exposed in this way.

★ Starred sections or subsections may be omitted, if the reader so wishes, as they are not required later in the book.

But this tight binding usually does not extend very far because inter-molecular forces are of short range and the forces acting on a molecule outside the solid are determined by the nature of the outermost layers. After the surface of the glass has been covered with the first one or two layers of vapour molecules, further vapour molecules approaching the surface experience an attraction which is almost the same as if the entire tube were composed of these molecules. Its surface energy per unit area is practically equal to that of the liquid from which the vapour was produced.

When a tube is first dipped into a liquid it has not risen in the tube and the surface is flat. Let us find the height to which it rises by calculating the change of potential energy when the tube is filled to an arbitrary height h and then let us write down the condition that the potential energy should be a minimum; this determines the equilibrium value of h. Let the liquid have density ρ and surface tension γ and let the radius of the tube be r. We can consider the tube to be filled in the following stages. We imagine a volume of liquid to be removed from the flat surface, just enough to fill the tube to the height h. It must have volume $\pi r^2 h$ and surface area $2\pi rh$ (Fig. 3.8). To remove it from the rest of the liquid an amount of energy equal to $(2\pi rh)\gamma$ must be expended. Then we imagine this liquid to be changed to a cylindrical shape (which requires no change of surface area and hence no expenditure of energy). When it is raised vertically, its potential energy due to its weight is increased to $(\pi r^2 h\rho)gh/2$ since its mass is $\pi r^2 h\rho$ and the height of its centre of mass is $h/2$. Finally the liquid can be imagined to be put inside the tube. A surface area of the inside of the tube equal to $(2\pi rh)$ is covered and an equal area of the surface of the cylinder of liquid is also covered. Thus, since both surfaces have surface energy γ per unit area, the surface energy is reduced by $4\pi rh\gamma$.

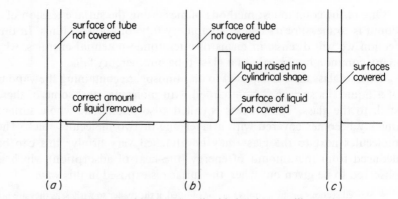

Fig. 3.8. Energy changes when liquid rises in a capillary tube.

Thus the total increase of potential energy when the liquid rises to height h is

$$U = 2\pi rh\gamma + \pi r^2 h^2 \rho g - 4\pi rh\gamma$$

$$= \pi r^2 h^2 \rho g - 2\pi rh\gamma.$$

This equation shows that the reason the liquid rises in the tube is that *the surface energy of the interior of the tube is reduced.*

The condition that U should be a minimum is that

$$\frac{\mathrm{d}U}{\mathrm{d}h} = 0,$$

that is

$$\pi r^2 h\rho g - 2\pi r\gamma = 0$$

$$h = \frac{2\gamma}{r\rho g}. \tag{3.8}$$

For carbon tetrachloride in a tube of 1 mm bore, h is equal to 6.5 mm since $\gamma = 26$ dyn/cm and $\rho = 1.6$ g/cm^3.

This discussion should be valid for any liquid whose molecules are adsorbed on to the surface of the tube. For mercury in glass, however, conditions are very different. Under normal conditions mercury does not adhere to glass. Droplets of this liquid simply run off a glass surface.

It should be noted however that under very special conditions mercury can be made to stick to glass, but that even then the adhesion is very weak. The effect is sometimes observed in McLeod gauges used to measure pressures in high vacuum systems. McLeod gauges have two limbs containing mercury, one open and the other closed. When the pressure is being measured in a system at extremely low pressure the mercury is pushed right up to the closed end of the tube and fills it completely. As part of the measuring procedure, the mercury in the other limb is then lowered and under normal conditions the mercury in the closed limb also falls, so that the two menisci keep at practically the same height as one another. But when conditions are exceptionally clean (the glass surfaces have been heated and the system has been evacuated for a considerable time), it is occasionally observed that the mercury in the closed limb does not fall but remains stuck to the glass. If the difference of heights is h, the mercury at the top of the closed limb is under a *tension* of ρgh dyn/cm^2, where ρ is the density of the mercury, and this tension must be resisted by the adhesion to the glass. As the level is further lowered, the mercury in the closed limb suddenly falls when the adhesion is broken.

Level differences up to about 10 cm are sometimes seen, corresponding to tensions of about 10^5 dyn/cm². We can use this fact to estimate the energy required to separate 1 cm² of mercury from glass. Let us assume that once we have separated them by a distance of one or two atomic diameters, the force becomes very small. The work required to do this is equal to the tension times the distance, which is 10^5 dyn/cm² multiplied by 10^{-8} cm, that is 10^{-3} erg/cm² for a range of 1 Å; for a range of 10 Å it is 10^{-2} erg/cm². Interpreting this in terms of interatomic potential energies as we did in section 3.4, the depth of the well must be of the order of 10^{-5} or 10^{-6} eV (using $\mathscr{N}$ of the order of 10^{15} per cm² and n about 10). This is a very small figure compared with quite weak van der Waals energies. Of course, the mercury glass bond always breaks at its *weakest* place and the *average* energy of adhesion must be rather larger than we have estimated. But allowing for this, it seems safe to assume that even under the most favourable conditions mercury adheres only weakly to glass.

This means that when a glass surface is covered with mercury, the surface energy decreases at best by very little. Following the previous analysis, the condition for minimum total potential energy is that the liquid should be depressed inside a tube, which is what is observed.

The important point about capillarity experiments is that they measure the interactions between molecules of the liquid and those of the surface of the tube. In order to measure the surface tension of the liquid alone, other methods have to be used.

3.4.4 Speed of ripples over a liquid surface

One interesting method of measuring the surface tension of a liquid is to find the speed of propagation of ripples across its surface.

When a wave is travelling across a surface and the wave profile is sinusoidal, the area is greater than when the surface is plane. The surface energy is therefore increased (Fig. 3.9). This effect—which leads to a finite speed of propagation of the waves—is the one we are interested in. But at the same time, the weights of the parts of the wave which are displaced upwards and downwards also increase the potential energy of the system and this affects the speed of propagation too. It can be shown that for waves of wavelength Λ and frequency f, the speed defined by

$$c = f\Lambda$$

is given by

$$c = \sqrt{\left(\frac{\Lambda g}{2\pi} + \frac{2\pi\gamma}{\rho\Lambda} \right)}, \qquad (3.9)$$

where γ is the surface tension of the liquid whose density is ρ, and g is the acceleration due to gravity.

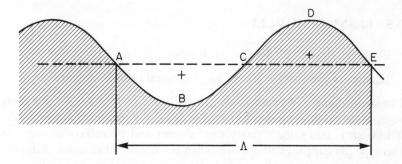

Fig. 3.9. Sinusoidal wave on a liquid. The perimeter ABCDE is longer than the undisturbed distance ACE (equal to the wavelength Λ) so that the surface energy is increased.

It follows from the dependence of the two terms on Λ and Λ^{-1} respectively that when the wavelength is very great the second term under the square root sign is small; the speed is then controlled by gravity alone and is not affected by surface tension. But when the wavelength is small, the gravity term becomes small and the speed is dominated by the surface tension term. For these short wavelength ripples,

$$c \approx \sqrt{\frac{2\pi\gamma}{\rho\Lambda}},$$

so that they travel faster the shorter their wavelength.

The method consists of generating ripples usually at audio frequencies, on the liquid in a tank. The frequency must be known and the wavelength is usually measured either by stroboscopic photography or by setting up a stationary wave pattern and measuring the distance between nodes. The liquid must be of sufficient depth for the bottom of the tank to have no influence on the waves. The disturbance does not in fact penetrate very deeply into the liquid; the effective mass which takes part in the motion is only about one-tenth of a wavelength deep, to be precise $\Lambda/4\pi$. The amplitude of the disturbance at a depth of one wavelength is negligible and this gives a criterion for the depth of liquid to use. The analysis of the motion is set as a problem at the end of the chapter. It depends on the fact that if a system is disturbed and its potential energy is proportional to the square of the displacement, the motion is periodic in time. This topic is discussed in section 3.6.

3.5 ELASTIC MODULI

Elastic moduli are all defined by equations of the type

(change of pressure) = (modulus)(fractional change of dimensions).

For small changes of dimensions, usually less than 1 % or 0.1 %, the body regains its original shape and size when the forces are removed; the behaviour is said to be reversible (or elastic), and we will concentrate on this type of change. Furthermore, when the fractional changes of dimensions are extremely small—a factor 10 or so smaller than the limit where elastic behaviour ceases—the changes of dimensions are quite accurately proportional to the pressures. Thus over these very limited ranges, the elastic modulus is a constant for the material. (We will discuss the breakdown of proportionality in section 3.7.1. and nonelastic behaviour in section 9.1.1.)

Stretching and twisting by different geometrical arrangements of forces, and compression by uniform 'hydrostatic' pressures are different ways of producing deformations and correspondingly we define Young's modulus, the rigidity or shear modulus and the bulk modulus as in Fig. 3.10.

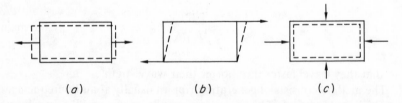

(a)   (b)   (c)

Fig. 3.10. Systems of forces and deformations defining elastic moduli. (a) linear tension producing extension, related by Young's modulus, (b) tangential forces producing an angle of shear, related by the rigidity and (c) hydrostatic pressure producing a change of volume, related by the bulk modulus.

For many practical purposes, it is necessary to emphasize the *differences* between the three moduli but here we will concentrate on their *similarities*. For any one substance they are of the same *order of magnitude*. Usually the bulk modulus and Young's modulus are almost equal and the rigidity modulus is a factor 2 or 3 smaller. This can be seen in the Table. But it must always be remembered that for liquids, the rigidity is zero (section 2.2.3).

Material	Young's mod. dyn/cm^2	Rigidity dyn/cm^2	Bulk mod. dyn/cm^2
Solid argon	7.0×10^9	3.0×10^9	6.0×10^9
Sodium chloride	4.0×10^{11}	1.3×10^{11}	2.5×10^{11}
Steel	2.0×10^{12}	0.8×10^{12}	1.8×10^{12}

We will take the bulk modulus as the typical elastic parameter, because it is the easiest to calculate.

The bulk modulus K of a material, the reciprocal of the compressibility, is defined by

$$K = -V \left(\frac{dP}{dV} \right) \tag{3.10}$$

where V is the volume, which is decreased when a pressure P is exerted uniformly in all directions. Usually, it is assumed that the temperature is kept constant during the compression. K can be measured directly by exerting a known pressure and measuring the change of volume—a whole technology has grown up for producing enormous pressures without the substance leaking past the piston which compresses it. Usually the main source of error is due to the non-uniformity of the forces acting in different directions. Alternatively, the speed of propagation of sound waves through a material can be found. This depends on the compressibility (as mentioned in section 3.6.1.), though a number of corrections have to be applied if exact values of K are required.

3.5.1   Bulk modulus and the $\mathscr{V}(r)$ curve

A compressed body can do work if the pressure is released; thus a compressed body has potential energy. This energy is given by a term of the type (force) × (distance) or (pressure) × (volume)—this is shown explicitly for a gas in Fig. 4.2 but it holds for any body. Notice that the energy E increases when the volume decreases so that,

$$dE = -P \, dV.$$

Hence we can write

$$P = -\frac{dE}{dV}. \tag{3.11}$$

This is an important relation. Pressure can usually be interpreted as an energy per unit volume, an energy density. However, one must be careful about this expression; it assumes that no heat flows in during the process of compression, or in thermodynamic language the compression must be adiabatic. This conflicts with the usual definition of bulk modulus given above. But thermal effects become small at low temperatures, so once again our estimates will become better the lower the temperature.

Substituting (3.11) in (3.10):

$$K = V\left(\frac{d^2E}{dV^2}\right)$$

This expression is in macroscopic terms—that is, the variables are E, the energy of the whole *block* of material, and V its volume. We wish to express these in terms of $\mathscr{V}(r)$ the potential energy of a pair of *molecules* and r the separation between two *molecules*. We do this as follows.

If we can express V in terms of r, we can express d^2E/dV^2 in terms of r. In general,

$$\frac{dE}{dV} = \frac{dr}{dV} \cdot \frac{dE}{dr},$$

$$\frac{d^2E}{dV^2} = \frac{d}{dV}\left(\frac{dE}{dr} \cdot \frac{dr}{dV}\right) = \frac{d^2E}{dr^2} \cdot \left(\frac{dr}{dV}\right)^2 + \frac{dE}{dr} \cdot \frac{d^2r}{dV^2}. \tag{3.12}$$

This holds whatever the relation between V and r. In particular, if we regard the molecules as little cubes.

$$V = Nr^3,$$

where N is the number of molecules in the block and r the distance between two neighbours. Therefore

$$\frac{dV}{dr} = 3Nr^2.$$

Further, if deformations are small and we take only nearest neighbour interactions into account, we can say that to a good approximation $r = a_0$, where a_0 is the separation for static equilibrium where the potential energy is a minimum, i.e. $dE/dr = 0$. Thus the second term on the right hand side of (3.12) is zero, and

$$\frac{d^2 E}{dV^2} = \left(\frac{d^2 E}{dr^2}\right)_{r=a_0} \bigg/ 9N^2 a_0^4,$$

$$K = \left(\frac{d^2 E}{dr^2}\right)_{r=a_0} \bigg/ 9Na_0. \tag{3.13}$$

Finally we must relate E, the energy of the *block* of material, to the energy of the individual *molecules*. As already emphasized, we will assume that their kinetic energy is small compared with their potential energy due to the intermolecular forces, which is equivalent to saying that we consider only low temperatures. We therefore relate E to $\mathscr{V}(r)$. Taking only nearest neighbour interactions into account, we consider $\frac{1}{2}N\mathfrak{n}$ nearest neighbour pairs; then

$$E = \tfrac{1}{2}N\mathfrak{n}\mathscr{V}(r),$$

and

$$K = \mathfrak{n}\left(\frac{d^2 \mathscr{V}(r)}{dr^2}\right)_{r=a_0} \bigg/ 18a_0. \tag{3.14}$$

This expression holds whatever the form of $\mathscr{V}(r)$. Let us assume the 6–12 potential

$$\mathscr{V}(r) = \varepsilon\left[\left(\frac{a_0}{r}\right)^{12} - 2\left(\frac{a_0}{r}\right)^6\right].$$

Then

$$\left(\frac{d^2 \mathscr{V}(r)}{dr^2}\right)_{r=a_0} = \frac{72\varepsilon}{a_0^2} \tag{3.15}$$

whence

$$K = \frac{4\mathfrak{n}\varepsilon}{a_0^3}$$

$$= \frac{4N\mathfrak{n}\varepsilon}{Na_0^3} = \frac{8L_0}{V_0}, \tag{3.16}$$

where V_0 is the molar volume and L_0 the binding energy per mole. This is a rather extraordinary relation. It predicts that, for solids which are bound together by van der Waals forces so that the 6–12 potential is a good description of the interatomic potential, the bulk modulus is 8 times the binding energy per unit volume. Implicitly, we are referring to low temperatures. This relation allows us to predict the *order of magnitude* of any elastic constant, if we know the latent heat of *evaporation* or the surface tension. The factor 8 depends on the assumption of the 6–12 potential but one expects a similar sort of factor for any molecular solid or liquid.

3.5.2 Comparison of bulk modulus and latent heat data

As in sections 3.3.1 and 3.4.2, we will use liquid nitrogen and carbon tetrachloride as typical molecular liquids. For liquid nitrogen, $8L_0/V_0 = (8 \times 6 \times 10^{10}/35)$ erg/cm^3 $= 1.4 \times 10^{10}$ erg/cm^3. The measured bulk modulus of the *solid* at about the same temperature is 1.26×10^{10} dyn/cm^2. The agreement is excellent. For CCl_4, at room temperature $8L_0/V_0 = 8 \times 3.2 \times 10^{11}/96 = 2.5 \times 10^{10}$ erg/cm^3. The measured bulk modulus is 1.1×10^{10} dyn/cm^2. The agreement within a factor of 2 must be considered good.

3.6   VIBRATIONS IN CRYSTALS: SIMPLE HARMONIC MOTION

We have already seen that molecules in a solid or liquid are vibrating about their mean positions (sections 2.2.3, 2.2.4). The purpose of the present discussion is to show how the frequency of vibration can be estimated, knowing the $\mathscr{V}(r)$ curve.

Firstly we must establish some relations about simple harmonic motion. Consider a system subject to a restoring force directed towards an origin and proportional to the displacement x. With the sign convention of section 3.2,

$$F = -\alpha x.$$

Such a system executes simple harmonic motion of frequency

$$\nu = \frac{1}{2\pi}\sqrt{\frac{\alpha}{m}},$$

where m is the mass in motion. This well known result can be recast in terms of potential energies. We are at liberty to take the zero of potential energy anywhere we wish. In this case we will choose it at $x = 0$, and

$$\mathscr{V} = \tfrac{1}{2}\alpha x^2. \tag{3.17}$$

The system is said to be in a parabolic potential well (Fig. 3.11(a)).

We can define the *curvature* at any point of a curve as the reciprocal of the *radius* of curvature. It is shown in mathematics texts that for the curve $y = f(x)$,

$$\text{curvature} = \frac{d^2y/dx^2}{\{1 + (dy/dx)^2\}^{3/2}}. \tag{3.18}$$

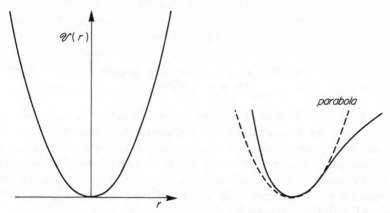

Fig. 3.11. (a) Parabolic potential well. (b) A curved potential well which is roughly parabolic at the bottom.

If the curve passes through a minimum then dy/dx is zero there and the curvature *at the minimum* is simply d^2y/dx^2. Thus for a parabolic potential well $\mathscr{V} = \frac{1}{2}\alpha x^2$, α is the curvature at the bottom of the well.

Finally we can say that *any* reasonable curve which goes through a minimum is not very different from a parabola in the vicinity of the minimum, Fig. 3.11(b).

Gathering these results together, we can say that a system which has a $\mathscr{V}(r)$ curve of the usual interatomic type will come to rest in static equilibrium at a separation a_0 at the minimum. But if it is displaced slightly, it will undergo simple harmonic oscillations of frequency

$$v = \frac{1}{2\pi}\left[\frac{(d^2\mathscr{V}/dr^2)_{r=a_0}}{m}\right]^{1/2}. \tag{3.19}$$

3.6.1   Einstein frequency

Imagine now a solid in which all the molecules are fixed at their equilibrium positions in the perfect crystal lattice, except one which is free to vibrate. As a first approximation, dissect this one out of the lattice together with two neighbours, one on either side, and assume that the vibration takes place along the line joining them, Fig. 3.12. When the molecule moves to the left it goes nearer to one neighbour and further away from the other. To the approximation that the potential well near the minimum is a symmetrical parabola, the change of potential energy is *twice* that due to one neighbour. So, using the result of Eq. (3.15),

$$\left(\frac{d^2\mathscr{V}}{dr^2}\right)_{r=a_0} = \frac{144\varepsilon}{a_0^2} \quad \text{(2 neighbours in line)}.$$

Fig. 3.12. Linear vibrations of a molecule with
two nearest neighbours.

To a better approximation, imagine that we dissect out the one molecule
surrounded by n nearest neighbours distributed uniformly over the
surface of a sphere: these neighbours are now at all angles to the direction
of vibration. The potential well is of depth $n\varepsilon$ and is a parabolic function of
radial distance r. But to calculate the change of potential energy with
displacement along a certain direction we have to average a factor $\cos^2 \theta$
over all directions. It emerges that

$$\left(\frac{d^2 \mathscr{V}}{dr^2}\right)_{r=a_0} = \frac{72n\varepsilon}{3a_0^2} \qquad \text{(n neighbours spherically disposed)}$$

like the previous expression but with the factor 3 coming from the averaging. Thus the frequency

$$\nu_E = \frac{1}{2\pi}\sqrt{\frac{24n\varepsilon}{ma_0^2}} \tag{3.20}$$

where m is the mass of one molecule. It is called the Einstein frequency.
This is a very rough estimate of the frequency of vibration, because of
course *all* the molecules are vibrating at once and the potential energy
of one molecule depends not only on its own position but on its neighbours'.
Whereas this analysis implies that all molecules are vibrating at a single
frequency, in a real solid many frequencies are present. But the order of
magnitude is significant. It is no coincidence that the Einstein frequency
and the bulk modulus both depend on $(d^2 \mathscr{V}/dr^2)_{r=a_0}$. The connection is
that the speed of sound is given by

$$\text{speed} = \sqrt{\left(\frac{\text{bulk modulus}}{\text{density}}\right)} \tag{3.21}$$

and using Eqs. (3.16) and (3.20) it can be verified that the Einstein frequency
is the frequency of sound waves whose wavelength is about twice the
intermolecular spacing. A full understanding of this result depends however on a study of the propagation of waves through lattices of points
rather than continuous media.

3.6.2 Estimation of Einstein frequency

From the data of 3.3.1 and 3.4.2, the frequency of molecular vibrations in
both liquid nitrogen and carbon tetrachloride is of the order of 10^{12} c/s.

This has no immediate interest for us though later we shall see that it has important consequences for the thermal properties, notably the specific heat, of these substances (5.4.4).

3.6.3   Experimental data for argon

Argon is a rare gas which liquefies at about the same temperature at which air liquefies, 80°K. The molecule is a single spherical atom. Interatomic attractions are purely van der Waals forces. The solid has a close packed structure. It is therefore an 'ideal' molecular crystal and has been extensively studied down to very low temperatures. Experimental data are given in Fig. 3.13; suggestions for analysing them are given in a problem at the end of the chapter.

One interesting use to which these data can be put is to verify experimentally that the 6–12 potential is a good representation. If we use the generalized p–q form of the interatomic potential energy, Eq. (3.3), then it can be verified that

$$\left(\frac{\mathrm{d}^2 \mathscr{V}}{\mathrm{d}r^2}\right)_{r=a_0} = \frac{pq\varepsilon}{a_0^2} \tag{3.22}$$

Thus comparing the bulk modulus with the binding energy L_0 per unit volume, as in equation (3.16), the product pq can be measured. It will be found to be about 64, which is surprisingly close to 72.

The sublimation energy has only been measured down to 70°K and this makes exact extrapolation to 0°K difficult. However, specific heats have been measured down to very low temperatures and we can then use the following energy cycle to find L_0. This uses quantities which have not yet been defined but is given here for completeness. (i) Start with the solid at $T = 0$°K and vaporize it; the energy required is L_0. (ii) Heat the gas to 83°K; to present accuracy it is a perfect gas of specific heat $\frac{5}{2}$R, where R is the gas constant = 8.31 J/deg. (iii) Condense the gas to solid at 83°K— the energy released can be read off the graph. (iv) Cool to near 0°K—the energy extracted, deduced from specific heat measurements at these low temperatures, is 165 J/mol. The argon is now back in its initial condition and from the energy balance L_0 can be calculated.

3.7   METALS

Metals consist of arrays of positive ions permeated by an atmosphere of free electrons (section 2.1.4). Each ion carries a positive charge and the electrostatic coulomb repulsions between the array of like charges would be enormous; the electrons, all negatively charged, would also repel one another equally strongly. Thus at first sight we would not expect metal to cohere but to fly apart. But in fact any small region in the metal tends to be electrically neutral and the electrons tend to concentrate between the ions so that their negative charges cancel out or screen the effect of the positive ions on one another.

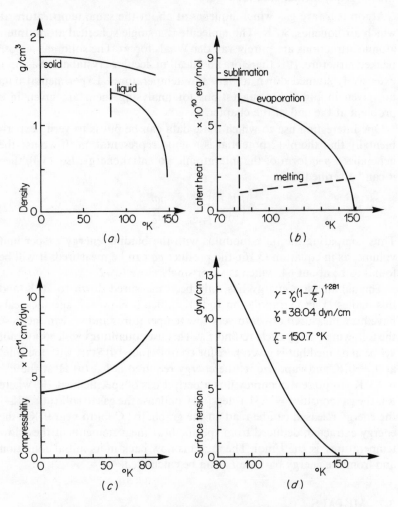

Fig. 3.13. Data for Argon. Atomic weight = 40. Sources of data : (a) *Densities* —Dobbs and Jones, *Rept. Prog. Phys.* **20**, 516 (1957); Mathias, Onnes and Crommelin, *Leiden Comm.* **131a** (1912). (b) *Latent heats*—computed from densities and vapour pressure measurements of Clark, Din and Robb, Michels, Wassenaar and Zwietering, *Physica* **17**, 876 (1951). (c) *Compressibility*—Dobbs and Jones, as above. (d) *Surface tension*—Stansfield, *Proc. Phys. Soc.* **72**, 854 (1958).

The problem of calculating the potential energy of an ion in the lattice is an extremely difficult one because the electrons are mobile and can redistribute themselves if the mean distance between ions is changed. Nevertheless we can make some general statements about the shape of the interionic potential energy curve.

First, there must be a minimum in the curve because the metal coheres and energy equal in magnitude to a binding energy is needed to evaporate it. We also observe that a metal resists great compression and we interpret this to mean that when the ions themselves begin to overlap the potential energy increases very rapidly, in much the same way as in a molecular solid. Thus the $\mathscr{V}(r)$ curve must resemble Fig. 3.4 near the minimum and at small values of r the curve must rise very steeply. As we have seen (section 3.2) we can use almost any rapidly increasing algebraic function of r to represent this.

Next we must discuss the $\mathscr{V}(r)$ curve on the other side of the minimum. A metal also resists being stretched or expanded so that the curve must rise in a similar sort of way. Now when we dealt with molecular solids we represented the potential energy of the van der Waals attractions between neutral molecules by an r^{-6} law; there are sound theoretical reasons for doing this. By contrast, there is *no* simple expression of this type to represent the subtle interplay of attractions and repulsions between the ions and the mobile electrons. Nevertheless the screening effect causes the interionic forces to be of *short range*. Quite arbitrarily we will therefore adopt an r^{-6} law for the potential energy at large r in metals also. The justification for this procedure is that it allows us to reach results which have the right order of magnitude so that, arguing in reverse, we can say that there must be a fairly strong resemblance between the real $\mathscr{V}(r)$ curve in metals, and Fig. 3.4. We will therefore use the Lennard–Jones 6–12 potential energy for metals also. But it must be clearly understood that for metals, in contrast to molecular crystals, it is only a crude approximation having no theoretical justification.

Some data for potassium and mercury are collected in the Table below. The coordination number in metals is always high, about 10 or 12. From the surface tension of the liquid (potassium at high temperature, mercury at room temperature) and also from the latent heat the depth ε of the potential well can be calculated. The two estimates agree tolerably well. The value of ε is comparable with that for molecular solids.

The bulk modulus does not agree very well with the ratio $8L_0/V_0$ (eight times the binding energy divided by the molar volume; see section 3.5.1). For potassium there is a discrepancy by a factor of 5 and for other metals it can be 10. The failure of this rather sensitive test shows that a 6–12 potential is not a good representation of the potential for some

	Potassium	Mercury
Atomic weight	39	200
Density (g/cm^3)	0.86	14
Binding energy L_0 (erg/g atom)	11×10^{11}	7.8×10^{11}
Surface tension of liquid (dyn/cm)	364	465
Bulk modulus (dyn/cm^2)	0.4×10^{11}	2.7×10^{11}
Atomic volume (cm^3)	45	14
Diameter of ion (Å)	4.2	2.8
ε from L_0 (eV)	0.19	0.16
ε from surface tension (eV)	0.14	0.09
$8L_0/V_0$ Eq. (3.16) (erg/cm^3)	2×10^{11}	4.5×10^{11}
Einstein frequency (c/s)	$\sim 10^{13}$	$\sim 10^{13}$

metals. If instead of 6 and 12 for the indices we use p and q, Eq. (3.3), then following through the calculation suggested in section 3.6.3 it is not difficult to show that the compressibility should be given by $pqL_0/9V_0$ ($8L_0/V_0$ is a particular case when $pq = 72$). Presumably the data for some metals mean that the product pq is sometimes a good deal smaller than 72; the repulsions might vary more slowly than r^{-12} or the attractive part of the interatomic potential might be of longer range than r^{-6}.

3.7.1   Departures from Hooke's law

In section 3.5.1 we limited the consideration of the elastic moduli of a solid to small departures from the position of minimum potential energy. The changes of dimensions and the pressures acting on the solid were both assumed to be small; under these conditions the deformation is proportional to the pressure and the substance is said to obey Hooke's law. Conditions like these are the ones usually encountered.

Now it must be emphasized that a solid may be perfectly elastic (in the sense that the body regains its original shape and size when all the forces are removed) and yet it may not obey Hooke's law. Indeed strictly speaking for any finite deformation, Hooke's law *cannot* hold. If from experiment the deformation is proportional to the pressure, this merely means that the deformation has not been measured accurately enough.

We can calculate the relation between pressure and deformation without imposing the condition that we are always near the minimum of the potential energy curve. Recapitulating some of the equations of section 3.5.1: pressure $P = -(dE/dV)$ where $E = \frac{1}{2}Nn\mathscr{V}$ and $\mathscr{V}$ is the pair potential. Therefore $P = -\frac{1}{2}Nn(d\mathscr{V}/dV)$. Though it is not a very good approximation, we will use the 6–12 potential for metals. It is convenient to write it in the generalized form of Eq. (3.3):

$$\mathscr{V} = 12\varepsilon\left\{\frac{1}{12}\left(\frac{a_0}{r}\right)^{12} - \frac{1}{6}\left(\frac{a_0}{r}\right)^6\right\}$$

and then to rewrite it in terms of volumes. Putting $V_0 = Na_0^3$ for the initial volume and $V = Nr^3$ for the volume under pressure:

$$\mathscr{V} = \varepsilon\left\{\left(\frac{V_0}{V}\right)^4 - 2\left(\frac{V_0}{V}\right)^2\right\}.$$

Substituting this in the equation for the pressure

$$P = -\tfrac{1}{2}Nn\left(\frac{d\mathscr{V}}{dV}\right) = \tfrac{1}{2}Nn\varepsilon\left\{\left(\frac{4V_0^4}{V^5}\right) - \left(\frac{4V_0^2}{V^3}\right)\right\}.$$

We can write this more elegantly by using the fact that the compressibility at small pressures is given by $K = 4Nn\varepsilon/V_0$. Therefore

$$P = \tfrac{1}{2}K\left\{\left(\frac{V_0}{V}\right)^5 - \left(\frac{V_0}{V}\right)^3\right\}. \tag{3.23}$$

Further, it is common practice to use the fractional change of volume $(V - V_0)/V_0$ as a measure of the deformation; it is called the *strain* and is denoted by s. Thus

$$\frac{V_0}{V} = \frac{1}{1+s}. \tag{3.24}$$

Substituting this and expanding by the binomial theorem for small s,

$$P = -K(s - \tfrac{9}{2}s^2 + \tfrac{25}{2}s^3 - \cdots). \tag{3.25}$$

This relation between pressure and strain should be obeyed by any solid with a 6–12 potential—molecular solids like argon, as well as metals. But metals have been extensively studied and can be prepared as specimens capable of undergoing large deformations, and more data exist for them than for any other class of material, so we will concentrate on them.

Our expression suggests plotting the ratio P/K as a function of strain s. This is reasonable: the definition of bulk modulus, extrapolated naively, implies that a pressure equal to K would reduce the volume by a factor e, so we can take K as a unit of pressure. The curve is plotted in Fig. 3.14. Positive pressures, which cause compression, are plotted downwards and negative pressures (that is, tensions) are plotted upwards. Hooke's law then appears as a straight line at 45°. As we shall see shortly, the small range of strains (between $\pm4\%$) covered by the graph encompasses an extremely wide range of conditions, far outside any encountered in ordinary engineering.

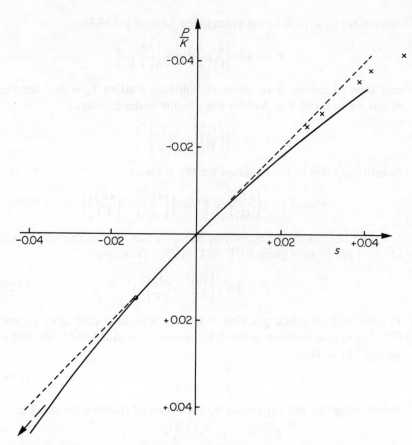

Fig. 3.14. Stress/strain curve predicted for a solid with a 6–12 potential (full line). Hooke's law is the dotted line at 45°. Positive strains, to the right, are extensions; negative strains are compressions. Pressures plotted *downwards*; tensions *upwards*; both measured in units of K. Crosses are measurements of the elongation of an iron whisker (Brenner, *J. Appl. Phys.* **27**, 1484 (1956)). Compression measurements on iron by Bridgman (*Proc. Am. Acad. Arts. Sci.* **77**, 187 (1949)) agree at $s = 0$ and $s = -0.017$, marked by an open circle; at intermediate stresses Bridgman gives the coefficient of s^2 as 6.1 where we have 4.5. Explosion waves show that for extremely large strains the points lie *above* our curve; this is suggested by the arrow. (Al'tschuler et al, *Soviet Phys. J.E.T.P.* **7**, 606 (1958)).

The trend of the curve is reasonable. When a solid is compressed, the repulsive forces come into play and a given pressure produces less strain than predicted by Hooke's law. The opposite holds for stretching.

Results for iron are shown at a number of points. These represent a range of techniques which is probably as wide as can be imagined. On the left the compression measurements are taken from experiments in which enormous pressures, of the order of 30,000 atmospheres, were generated hydraulically using thickwalled vessels with ingeniously designed pistons. These experiments showed that for small strains the coefficient of s^2 is 6.1, where our simple theory gives 4.5, so that the points lie below our curve; but then they move upwards and at about $s = -0.017$ the points lie on the predicted curve. Other experiments have also been done at far higher pressures, right off the graph—experiments in which high explosive was detonated on the face of a thick iron plate and the speed of the shock wave determined. Measurements at $s = -0.2$ and beyond show that the points lie above our predicted curve; the arrow attempts to indicate this. This result may mean that the real repulsive force cannot be represented by a simple power law like r^{-12}.

In contrast to these massive techniques, the other half of the curve represents experiments performed under a microscope. It is a fact that if iron or any other ordinary metal is strained beyond about 0.1%, it breaks. But it is possible to prepare 'whiskers', that is thin threads of the metal which—for reasons which will be described in section 9.4.2—are by comparison immensely strong. The one used in these experiments was 1.6×10^{-4} cm in diameter and a few millimetres long. It was stretched by applying a force equal to the weight of about 10 g and the extension measured. Of course, such an experiment measures Young's modulus, not the bulk modulus, but as we have stated in section 3.5, these are almost the same.

Our simple theory gives deviations from Hooke's law of the correct sign, and correct magnitude within a factor of 2, which is satisfactory.

3.8   IONIC CRYSTALS

Lattices like those of sodium chloride and lithium chloride consist of arrays of ions, each positively or negatively charged (Fig. 2.2). The forces and potentials between two ions therefore consist now of *three* components: the repulsive part, the van der Waals part (exactly as for neutral atoms) and in addition the electrostatic attraction or repulsion. This is given by Coulomb's law. If each ion carries a charge of magnitude e the potential energy $= +e^2/4\pi\varepsilon_0 r$ (like charges) or $-e^2/4\pi\varepsilon_0 r$ (unlike charges). The energy is measured in joules if e is expressed in Coulombs, r in metres and ε_0, the permittivity of free space is given by $4\pi\varepsilon_0 = 10^{-9}/9$ farads/metre. The signs follow the convention of Fig. 3.1, to express the direction of the attraction or repulsion. We summarize this as

$$\mathscr{V}(r)_{\text{Coulomb}} = \pm \frac{e^2}{4\pi\varepsilon_0 r} \quad \begin{cases} \text{Like charges} + \\ \text{Unlike charges} - \end{cases}$$

Thus the total potential energy between two ions is

$$\mathscr{V}(r) = \frac{\lambda}{r^p} - \frac{\mu}{r^6} \pm \frac{e^2}{4\pi\varepsilon_0 r},$$

where p is about 10 and λ and μ should be of the same order of magnitude as for neutral atoms.

We will now show that the r^{-6} term, the van der Waals attraction, is negligible compared with the coulomb potential and can be discarded. If the charge cloud were spherically symmetric, the Coulomb force and potential at an external point would be the same as if its charge were concentrated at the centre. This is a consequence of the inverse square law. Thus the potential energy between two ions cannot be very different from that between an ion and an electron the same distance away. But we know that to pull an electron out of an atom to make an ion, we have to do an amount of work (equal to the ionization energy) of the order of 10 eV, that is 10^{-11} erg. Therefore, since the diameter of an atom is of the same order as the interionic spacing in crystals, the potential energy of two ions must be of this order of magnitude. The van der Waals energies are of the order of 0.1 eV, one hundred times smaller.

In Fig. 3.15 we have attempted to plot the potential energy of two ions of like sign and unlike sign as a function of their separation. Also shown is

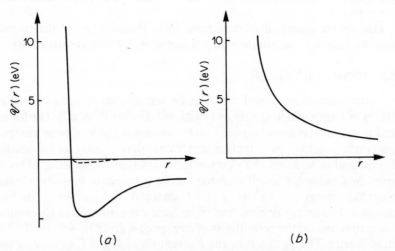

Fig. 3.15. (a) The potential energy of two ions of unlike charge as a function of the distance r between their centres. The dashed curve is the potential energy of two neutral atoms, Fig. 3.4, plotted on the same scale. (b) The potential energy of two ions of like charge, which is repulsive at all distances.

the potential energy of two neutral atoms, to the same scale. This emphasizes in an obvious way that van der Waals energies can be neglected for ions. We can therefore write for two ions:

$$\mathscr{V}(r) = \frac{\lambda}{r^p} \pm \frac{e^2}{4\pi\varepsilon_0 r} \begin{cases} \text{Like charges} + \\ \text{Unlike charges} - \end{cases} \qquad (3.26)$$

3.8.1   The binding energy of sodium chloride. The Madelung sum.

Figure 3.15 also emphasizes another point: that the Coulomb potential is long range (section 3.2). It is therefore no longer justifiable to deal only with nearest neighbour interactions. The Coulomb potential due to an ion can be felt far into the lattice, and another method of calculating binding energies is needed.

To start with, consider the *Coulomb energy only* of a *line* of ions, each of charge e alternately positive and negative as in Fig. 3.16, which has been dissected out of a lattice like that in Fig. 2.2.

Let the interionic spacing be r. Then the energy of *one* ion due to its two neighbours (which necessarily have the opposite sign to it) is $-2e^2/4\pi\varepsilon_0 r$. The next nearest neighbours, distance $2r$ away and necessarily having the same sign as the ion considered, give potential energy $+2e^2/4\pi\varepsilon_0 2r$; and so on.

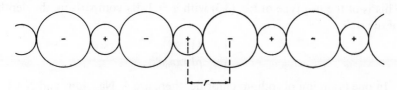

Fig. 3.16. A line of ions dissected out of an ionic lattice

Thus the potential energy of the single ion in the infinitely long line is

$$\frac{-2e^2}{4\pi\varepsilon_0 r}(1 - \tfrac{1}{2} + \tfrac{1}{3} - \tfrac{1}{4} + \cdots).$$

Note that

$$\log_e(1+x) = x - \frac{x^2}{2} + \frac{x^3}{3} \cdots$$

so that

$$1 - \tfrac{1}{2} + \tfrac{1}{3} - \tfrac{1}{4} \cdots = \log_e 2 = 0.69.$$

Therefore the potential energy of the single ion in the line is $-1.38e^2/4\pi\varepsilon_0 r$. This holds for any ion, positive or negative, anywhere in the line.

The constant which we have just worked out is called the Madelung constant α for a line of ions. The Madelung constant for the three-dimensional sodium chloride lattice has been calculated: it is 1.75, which is not very different from our 1.38. Indeed, all Madelung sums for simple lattices of ions of alternate sign are of the order of unity. They must be so, since the effect of a positive ion is to some extent cancelled out by the next nearest negative ion and so on. Thus although the Coulomb potential is long range, our calculations are greatly simplified by being able to say that the potential energy of an ion in a lattice is given by the energy of a nearest neighbour pair times the Madelung constant which is of the order of unity.

So far we have dealt with the Coulomb part of the potential. When it comes to the short range repulsions, the λ/r^p term in which p is about 10, we may guess that only nearest neighbours need be counted. The repulsive potential energy of a single ion is therefore $n\lambda/r^p$, where n is the coordination number.

The potential energy of a pair of ions in the crystal can therefore be conveniently written

$$\mathscr{V} = \frac{\alpha e^2}{4\pi\varepsilon_0 a_0}\left[\frac{1}{p}\left(\frac{a_0}{r}\right)^p - \left(\frac{a_0}{r}\right)\right] \tag{3.27}$$

This is of the p–q type of Eq. (3.3) with $q = 1$. By comparison, the depth of the well

$$\varepsilon = \left(1 - \frac{1}{p}\right)\frac{\alpha e^2}{4\pi\varepsilon_0 a_0}.$$

In one gram ion of sodium chloride, there are N Na^+ ions and N Cl^- ions, that is there are N pairs of ions. Thus when the interatomic spacing is r, the energy is $N\mathscr{V}$ per gram ion; when it is a_0, the energy is $-N\varepsilon$. This gives the binding energy of the substance.

Now the index p of the repulsive potential is about 10. So to about 10% accuracy, we may say that the binding energy of an ionic crystal is equal to $e^2/4\pi\varepsilon_0 a_0$ the potential energy of one pair of adjacent ions times the Madelung constant times Avogadro's number. The binding energy is dominated by the Coulomb energy.

The binding energy for sodium chloride has been determined as 763 kJ/mol. The interionic spacing has been determined by X-ray analysis as 2.8 Å that is 2.8×10^{-10} m. The charge e, on each ion, is one electron charge, 1.6×10^{-19} C. Therefore

$$\frac{N\alpha e^2}{4\pi\varepsilon_0 a_0} = 860 \text{ kJ.}$$

If we diminish this by 10% to allow for the factor $(1 - 1/p)$, the binding energy agrees extremely well.

3.8.2   Elasticity of ionic crystals

We can calculate the bulk modulus of an ionic crystal using the same method as for molecular crystals. The volume occupied by N pairs of positive and negative ions is

$$V_0 = 2Na_0^3$$

and equation (3.13) becomes

$$K = \frac{(d^2E/dr^2)_{r=a_0}}{18Na_0} = \frac{N(d^2\mathscr{V}/dr^2)_{r=a_0}}{18Na_0},$$

where $\mathscr{V}$ is the pair potential energy. The easiest way to evaluate the second differential is to quote the result of Eq. (3.22), that it is equal to $pq\varepsilon/a_0^2$, with $q = 1$. This leads straightforwardly to the result

$$K = \frac{pL_0}{9V_0},$$

where L_0 is the binding energy and V_0 the molar volume. This is a special case of the relation pointed out in 3.6.3, namely that if the indices in the interatomic potential are p and q, the bulk modulus depends on the product (pq); here $q = 1$, so we can measure p directly.

For sodium chloride, the molecular weight is 58.5 and the density $2.18\,\mathrm{g/cm^3}$ so that the molar volume V_0 is $27\,\mathrm{cm^3}$. The binding energy is $7.6 \times 10^{12}\,\mathrm{erg/mol}$, and the measured bulk modulus at low temperature is $3.0 \times 10^{11}\,\mathrm{dyn/cm^2}$. Substituting, p is found to be 9.4, which is reasonable.

The speed of sound, calculated from the bulk modulus and the density, is about $4 \times 10^5\,\mathrm{cm/s}$. The Einstein frequency ν_E, corresponding to a wavelength of about twice the interionic spacing, is therefore about $5 \times 10^{13}\,\mathrm{c/s}$. Since the propagation of such a wave means that ions (charges) are moving it can lead to the absorption of electromagnetic waves of this frequency. This frequency lies in the infrared.

PROBLEMS

3.1. For gravitational forces, which obey an inverse square law, a sphere of large radius behaves as if it were a point located at the centre. In this question, the object is to link up the definition of potential energy $\mathscr{V}$ with that of gravitational potential energy as usually defined in elementary treatments.

Write Newton's constant as G, the mass of the Earth M, the radius of the Earth a.

(a) Calculate the force on a mass m at a distance r from the centre of the Earth $(r > a)$. Take r radially outward, note that r can only be positive. Get the sign of this force correct.

(b) Calculate the potential energy, taking the value at $r = \infty$ to be zero.

(c) Draw a graph of this function between $r = a$ and $r = \infty$.

(d) Calculate the force on m at the surface of the Earth and by equating it to the weight mg, get an expression for g at the surface.

(e) Calculate the potential energy at $r = (a+h)$ where h is small compared with a, so that the square of h/a can be neglected. Show that the potential energy at $(a+h)$ is greater than that at a by an amount mgh. Mark this increase clearly on your graph.

3.2. Two small magnets are arranged as shown. The lower one is fixed, the upper one is restrained from moving horizontally, but is free to move vertically.

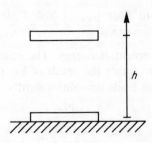

The force between them is a repulsion of magnitude $(2\mu_0 M^2/h^4)$ where M is the magnetic moment of each, μ_0 the permeability of vacuum and h is the distance apart of the two magnets.

(a) Write down the potential energy due to the repulsion.

(b) Draw a graph of this as a function of h.

(c) Draw a graph of the potential energy of the upper magnet as a function of h, due to the Earth's gravity.

(d) Draw a graph of the total potential energy as a function of h.

(e) Calculate h for static equilibrium and indicate this point on the graph.

(f) Using the methods of section 3.6, calculate the period of oscillation after the upper magnet is given a small displacement downwards and then released.

3.3. A sinusoidal wave $y = a \sin 2\pi x/\Lambda$, of wavelength Λ and amplitude a disturbs the surface of a liquid of density ρ and surface tension γ. The problem is to calculate the frequency and speed of propagation; see section 3.4.4 and Fig. 3.9.

(a) Calculate the potential energy of a whole wavelength, due to the up and down displacements of the weights of the two halves, as follows. Show that the displaced mass of the half wavelength between $x = 0$ and $x = \Lambda/2$ is $\rho a l \Lambda/\pi$, where l is the width of the wavefront. Prove that its centre of gravity is at height $\pi a/8$ and hence that the potential energy of a whole wavelength is $\frac{1}{4}\rho a^2 l \Lambda g$. Note: $\int \sin^2 \alpha \, d\alpha = \frac{1}{2}(\alpha - \sin \alpha \cos \alpha)$.

(b) Write down an integral expressing the length of perimeter of a whole wavelength of the sinewave. Assume a/Λ is small, expand the integrand by the binomial theorem and integrate it. Show that the increase of surface energy is $\pi^2 a^2 l \gamma/\Lambda$. The total potential energy is the sum of this and (a); note the proportionality to a^2.

(c) Assume (from comparison with a complete analysis) that the penetration depth is effectively $\Lambda/4\pi$ so that the mass in motion is effectively $\Lambda^2 \rho l/4\pi$.

Hence show that the frequency is

$$v = \frac{1}{\Lambda}\sqrt{\frac{\Lambda g}{2\pi} + \frac{2\pi\gamma}{\rho\Lambda}}.$$

(d) The (phase) velocity is $v\Lambda$. Sketch it as a function of Λ.

(e) For waves with $\Lambda = 1$ mm on liquid argon at $100°$K, show that surface tension contributes 30 times as much to the energy as the weight. Calculate the frequency and velocity. Design an experimental set-up to measure the surface tension.

3.4. In a rough demonstration experiment, liquid nitrogen was contained in a small thin-walled, spherical glass dewar with a narrow neck. This was connected to a gas-meter so as to measure the volume of nitrogen boiled off. The meter and the gas passing through it were at room temperature, 20°C. While 25 g of liquid evaporated, the meter registered 0.76 cu ft of gas (1 cu ft = 28.3 litres). What is the molecular weight of nitrogen? The dewar was known to hold 122 cc up to a mark on the narrow neck. It was estimated that about 1 cc was always occupied by bubbles. The flask was weighed empty and full up to the mark; the liquid weighed 98 g. What is the density? the molar volume? the diameter of a molecule? A capillary tube of internal *diameter* 0.55 mm was dipped into the liquid. The capillary rise was 8 mm. What is the surface tension, and depth ε of potential well?

3.5. Calculate some of the atomic constants of argon using the data of Fig. 3.13; see section 3.6.3.

(a) From the density, Fig. 3.13(a), calculate the molar volume V_0 and hence the diameter a_0 of an argon atom, assuming each atom to be a little cube.

(b) If the details of the argon crystal lattice are taken into account, see section 8.1.2, it can be shown that the molar volume is not Na_0^3 but $Na_0^3/\sqrt{2}$. Calculate a better value of a_0 than in (a) above.

(c) Estimate the depth ε of the potential well from the latent heat data of Fig. 3.13(b), taking the heat of sublimation at $70°$K to be a sufficiently good measure of L_0 (Eq. (3.6)).

(d) Extrapolate the heat of sublimation to $T = 0°$K using the energy cycle described in section 3.6.3 and calculate a better value of ε than in (c) above.

(e) Estimate ε from the surface tension data of Fig. 3.13(d), extrapolating γ to $T = 0°$K.

(f) Estimate the bulk modulus K at $T = 0°$K from the compressibility graph, Fig. 3.13(c). Compare this value with $8L_0/V_0$ (Eq. (3.16)).

(g) Check Eq. (3.22). (Refer if necessary to section 8.4.1.) Hence show that if the indices of the interatomic potential energy are p and q instead of 6 and 12, the bulk modulus is given by

$$K = \frac{pqL_0}{9V_0}.$$

Use your values of K, L_0 and V_0 to estimate the value of pq.

(h) Estimate the Einstein frequency v_E (Eq. (3.20)).

3.6. The molecules of a complicated organic molecule can be considered to be disc-shaped, of radius r cm and thickness $r/10$ cm. When two molecules are close together face to face (like two pennies one on top of the other) an energy E is required to separate them; when they are placed end to end (like two pennies touching one another, edge to edge) an energy $E/30$ is required. For these molecules, which are far from spherically symmetrical, it is no use quoting formulae for L_0 or γ (Eqs. (3.6), (3.7)) for spherical molecules; it is necessary to imagine how these molecules can be packed together and to tackle the problem *ab initio*.

(a) Estimate the molar heat of evaporation.

(b) Estimate the surface tension.

(c) Near the surface of a solid or liquid, will the molecules tend to be oriented (i) randomly, (ii) with their planes parallel to the surface, (iii) with their planes normal to the surface? Give reasons.

3.7. The crystal of sodium fluoride is a cubic structure in which alternate sites are occupied by Na^+ and F^- ions, each ion having 6 nearest neighbours. The Madelung constant is 1.75 when referred to the smallest Na^+–F^- separation, which will be denoted by a_0. The heat of formation of the crystal from its constituent ions is 900 kJ per mole of sodium fluoride. The density of crystalline sodium fluoride is 2.9 g cm^{-3}, and the atomic weights of sodium and fluorine are 23 and 19. Estimate a_0 and Avogadro's number from the data. State clearly any approximations you are using in your calculations.

3.8. The interaction between ions in sodium chloride can be described by their Coulomb interaction, plus a repulsive potential energy $A \exp(-r/\rho)$ acting between nearest neighbours only. (This exponential term is used in place of the r^{-12} term; A and ρ are constants). Obtain an expression for the lattice energy in terms of the nearest-neighbour separation a_0 and the Madelung constant α for sodium chloride. Given $a_0 = 2.8$ Å, $\alpha = 1.75$, and that the lattice energy is 763 kJ/mol, find ρ.

3.9. In a medium of dielectric constant K, the potential energy of two charges e_1 and e_2 separated by distance r is $e_1 e_2/4\pi\varepsilon_0 K r$.

Water H_2O, ethyl alcohol C_2H_5OH and ammonia NH_3 have dielectric constants of 80, 25 and 18 respectively at room temperature. These high values are due to the fact that the molecules are electric dipoles, with regions of + and − charge, which can be easily aligned by external fields; they can crowd round a charged particle with their oppositely-charged ends all pointing towards it and so screen the particle from its neighbours. H_2O and NH_3 are compact molecules, C_2H_5OH is relatively large.

(a) Use the data for the binding energy of sodium chloride, section 3.8.1, to calculate the binding energy of a single sodium chloride molecule.

(b) Explain qualitatively the fact that the solubility of sodium chloride in 100 g of solvent is 37 g in water, 0.07 g in alcohol, 3 g in liquid ammonia; the dissolved sodium chloride exists as Na^+ and Cl^- ions.

3.10. Calculate the Madelung constant for a line of dipoles, (a) all aligned in the same direction so that they repel one another with an r^{-4} force, and (b) alternately parallel and antiparallel so that nearest neighbours attract, next nearest neighbours repel and so on. Evaluate the series numerically or by looking up Riemann zeta functions in Dwight's 'Mathematical Tables' or 'Tables of Integrals'.

3.11. The molecular weight M, latent heat of evaporation L, and surface tension γ at 20°C, of certain liquids are given in the table. Make a crude estimate of the molecular diameters in each case stating what approximations are made.

	M	L J/gm	γ dyn/cm
Alcohol	46	856	22
Benzene	78	389	29
Mercury	200	272	475
Water	18	2,250	73

3.12. (*i*) Prove that for a substance whose interatomic potential energy is given by Eq. (3.3), the relation between pressure P (applied hydrostatically) and volume v is

$$\frac{P}{K} = \frac{3}{p-q}\left\{\left(\frac{v_0}{v}\right)^\alpha - \left(\frac{v_0}{v}\right)^\beta\right\},$$

where K is the bulk modulus at very small strains, v_0 is the initial volume, and α and β are equal to $(p+3)/3$ and $(q+3)/3$ respectively. Hence show that for small strains s less than about 0.1

$$P = -K\left(s - \frac{p+q+9}{6}s^2 + \cdots\right).$$

(*ii*) For an ionic crystal, p is about 11 and $q = 1$. Compare this simple theory with measurements by Bridgman on the compression of sodium chloride.

p Kg/cm^2	s	p Kg/cm^2	s
1×10^4	-0.038	4×10^4	-0.115
2	-0.068	6	-0.152
3	-0.093	8	-0.183
		10	-0.210

One method of deducing the bulk modulus for very small strains is to note that on almost any theory $P = -K(s - \gamma s^2)$ where γ is a constant, so that a graph of P/s against s should be a straight line from which K can be deduced. Use the measurements in the first table for this.

(*iii*) Convert the data of the next table into (P/K) against (v_0/v) raised to the appropriate powers, compare calculated and measured pressures and comment on the results.

3.13. What conclusions can you draw about the natural vibrational frequencies of diamond, iron and lead from the following data (Y, M and ρ are respectively: Young's modulus in dyn/cm^2, the atomic weight, and the density in g/cm^3):

	Y	M	ρ
Diamond	8.4×10^{12}	12	3.5
Iron	2.0×10^{12}	56	7.9
Lead	0.18×10^{12}	208	11.4

CHAPTER

Energy, temperature and the Boltzmann distribution

4.1 HEAT AND ENERGY

We will take it for granted that heat is a form of energy. This statement is based on the experiments of Joule, the majority of which were of the same basic pattern. Weights held on strings could descend and so provide mechanical energy to drive a mechanism whose motion was resisted by friction of some kind and which grew hotter as it was driven. The change in mechanical energy was measured by the loss in potential energy of the weights, and the quantity of heat produced was measured in terms of the rise in temperature of the apparatus and its heat capacity. The mechanisms were very varied: a dynamo dissipating its energy in a resistance, a perforated piston moving through viscous liquids, a conical bearing with friction between the rubbing surfaces, a system of paddles churning viscous liquids. In another investigation, air was compressed by a pump into a cylinder and the temperature rise was measured—here no frictional force was encountered during the compression of the gas, but work had to be done against the pressure it exerted. In all these experiments, the conversion factor relating the energy absorbed by the mechanism and the heat produced in it was the same within rough limits, $\pm 15\%$. Since the mechanisms were so diverse in type, it was unreasonable to suggest that this rough constancy could be a property of the substances or devices employed; it could only be explained if heat and energy were physically identical.

4.1.1   Ordered and random movements of molecules

It has already been mentioned that the molecules of any substance are in ceaseless, rapid motion. The molecules of a gas move in straight trajectories till they collide, the molecules of a solid are in vibration about their mean positions, those in a liquid vibrate and also slip through the holes in the structure. These motions are random, in the sense that movement in one direction is just as likely as movement in any other, and also in the sense that any molecule changes its speed many times per second so that if we were to follow all the details of the motion we would find that the kinetic energy went irregularly through all possible values, from zero up to some large value.

Consider a body which is big enough to contain a large number of molecules (though it might be of dimensions which are small on the ordinary scale) and let it be at rest. Then the total momentum of the molecules must be zero. If, however, we examined the momentum of one single molecule at any instant, we would almost certainly find it to be large. It is only by finding the vector sum of all the momenta or finding the average momentum of a very large number of molecules at any instant, or alternatively by finding the average momentum of one molecule over a long period of time, that we can come to any conclusion about the movement of the body as a whole. In the same way, we can consider a region of a body which is moving with a certain velocity. If we were to determine the instantaneous velocity of any one molecule we might well find that it was moving very fast in the opposite direction to the bulk motion, and we could deduce nothing about that bulk motion. But if we averaged the momentum of a large number of molecules in the region at one particular instant, or averaged the momentum of one molecule over a long time, we would detect a nett momentum, corresponding to the drifting of the body as a whole.

It is important to point out that the statements just made are in fact a little too dogmatic. If we found the vector sum of the momenta of all the molecules in a body, it would be unlikely to be *exactly* zero. It would fluctuate about the value zero. For the averaging to have any physical significance, these fluctuations must be relatively small. We will see later (in section 7.7.2) that this condition is satisfied if the number of particles in the assembly is large, and it will be assumed in the rest of this chapter that this is so.

We are thus led to distinguish between the random movements of the molecules of a body (which add up vectorially, and hence average out to zero) and the movement of the body as a whole. The random movements are superimposed on the bulk or ordered movement, and the two can only be separated by averaging. An example of ordered motion is the macroscopic movement of any body such as a ball. A less obvious example is the flow of a liquid, either streamlined or turbulent, where a bulk velocity can be defined over each small region of the fluid. Another example is provided by waves of compression or rarefaction passing through any medium, or torsional waves through a solid, when the velocity and displacement due to the wave can be defined. The passage of an electromagnetic wave through a transparent medium may also cause the atoms to vibrate in an ordered way. But in all these examples, the ordered movements have the random movements of the molecules superimposed on them.

4.1.2   Temperature and random motion

We can now extend the statement that heat is a form of energy and make it more precise, as follows: Temperature is a measure of the energy of the random motion of the molecules of a substance. (Temperature is not the only measure of this energy, but we will not pursue this fact here.) This statement is ultimately based on Joule's experiments. Later in this chapter we will show how the form of the relation can be deduced.

A molecule in general possesses both kinetic and potential energy. Sometimes the kinetic energy dominates the situation, as in a gas of low density where the potential energy arising from close collisions can often be neglected. On the other hand, a solid owes its regularity of structure to the potential energy of interaction of an ion with its neighbours and as the kinetic energy merely leads to vibrations of the ions about their mean positions, the random variation of the potential energy is important.

It must be emphasized that it is only the randomly varying contribution to the energy which is related to the temperature. Any ordered movements must be transformed away by a suitable choice of coordinate system. However, it is possible for ordered molecular movements to be converted into random ones. In fact, this is just what happens whenever mechanical energy is converted into heat. In Joule's experiments where liquids were stirred, the movements of the weights were converted by the mechanisms into bulk movements of the liquids—the surface waves, the eddies. But all the time, part of the energy of the ordered motion of the liquids was being converted or degraded by molecular collisions into random movements of the molecules and this caused the rise in temperature. In his experiments where gases were compressed by a piston, the ordered motion of the layer of gas being pushed back by the movement of the piston was transferred to a random movement of molecules throughout the gas so that the energy of the piston caused a rise in temperature. At the same time, the molecules of the piston were also heated.

4.1.3   Degradation of ordered into random motion

To see how this conversion or degradation is carried out, we consider first a simple example. Let a single molecule be projected into a layer of

gas (Fig. 4.1), with a velocity much higher than the random velocities of
the molecules of the gas. In principle, we can imagine that this energy
could be measured by some simple mechanical device, since only one
particle is involved. The molecule must soon undergo a collision in which
the laws of conservation of energy and momentum are obeyed. A proportion of its energy is transferred to the other molecule, both are deflected
and the first continues at a slightly lower velocity. It undergoes a large
number of such collisions, each time transferring some of its energy and
deflecting the molecules of the gas till finally it emerges with a smaller
kinetic energy than it started with; in principle this loss of energy can be
measured by mechanical means. Each molecule struck undergoes further
collisions, at different glancing angles, with changes of direction each time,
so that after quite a short time the excess of energy is carried to distant
parts of the gas and is shared between all the molecules as extra kinetic
energy of motion in all directions. The energy of the molecules is increased
by an amount equal to the difference between E and E', but in macroscopic terms we merely say that the temperature has increased.

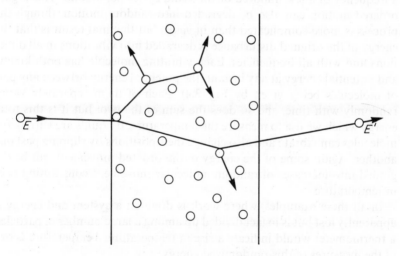

Fig. 4.1. Passage of a single fast molecule of initial energy E through a layer
of gas. It emerges with a lower energy E'.

We can now consider a gas contained inside a cylinder fitted with a
piston (Fig. 4.2). Let the area of cross section be A and the pressure of the
gas P, and imagine the piston to be moved inwards a distance dx. This
requires the expenditure of an amount of work $PA \cdot dx$. But each molecule

near the piston acquires an excess velocity and now plays the role of the single incident molecule in Fig. 4.1. Once again, the excess energy supplied by the moving piston is, after a large number of collisions, shared between all the molecules as a purely random movement, as a rise in temperature.

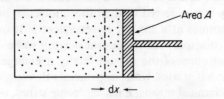

Fig. 4.2. Gas contained in a cylinder with piston. When the piston moves the ordered motion of the molecules near it is randomized by collisions.

If a solid is struck, the energy of deformation will at first probably be in the form of a compression wave travelling through it, so that it rings with a frequency of a few hundred or thousand cycles per second. This highly ordered motion can also be degraded into random motion, though the process is more complicated than in gases, but the final result is that the energy of the original disturbance is degraded into vibrations in all directions and with all frequencies. Each vibrating molecule has both kinetic and potential energy at any instant (the potential energy between any pair of molecules being given by Fig. 3.4). Each of them separately varies randomly with time, and so does the sum of the two, but it is this total energy which we use to measure the temperature. If liquids are stirred, the molecules can vibrate and also change their positions by slipping past one another. Again some of the energy of the ordered movement can be degraded into increases of random molecular movement constituting rises in temperature.

In all these examples, where work is done on a system and energy is apparently lost but is in fact divided up among a large number of particles, a thermometer would indicate a rise in temperature. Temperature is one of the measures of this randomized energy.

4.1.4 Macroscopic variables and statistical specifications

In practice, measuring instruments which we use for ordinary kinds of physical measurements are relatively massive, slowly-responding devices which are incapable of detecting the effect of individual molecules. Thus when we measure the pressure of a gas using an ordinary sort of pressure gauge, a mercury manometer or a diaphragm actuating a lever, the gauge measures a time-average of the pressure. It does not register the individual impacts of the molecules, which occur with extremely high frequency. Similarly, when we measure the density of a gas we normally use a volume whose dimensions are very large compared with the distances between molecules. Figure 2.3, which pictures the molecules in a gas unevenly distributed throughout a volume, implies that the density fluctuates from place to place, but these fluctuations are never detected by a measurement using a large volume. Thermometers, similarly, measure time-averages of molecular energies. It is in fact possible to detect fluctuations in pressure, or density or energy under the right conditions using appropriate detectors, but for the moment we will restrict our considerations to ordinary instruments which measure average values of molecular quantities.

The contrast between the complexity of the situation from the atomic point of view, and the simplicity of the large-scale measurements is a profound one. The one requires a knowledge of say 10^{23} positions and velocities, the other a single dial reading. It is our task to relate these points of view and it is suggested at once that the elaborate specification of all the positions and velocities of the constituent molecules is not only impossibly complicated, *but is actually irrelevant.*

We need only concern ourselves with those features of the assembly of molecules which allow us to calculate *averages* of molecular quantities. It turns out that the most we ever need to specify is the fraction of the total number of molecules to be found within certain limits of position or having speeds within a certain range. We need to know, for example, that 1 % of the molecules of a liquid of molecular weight 30 at 300°K have speeds between 10,000 and 10,320 cm/s; it is not necessary to enumerate which molecules they are at any instant nor exactly where they are to be found. Specifications of this kind are all that is necessary to relate the macroscopic properties to the motions of the molecules. One often meets rather similar statements about, say, the economic state of a nation. For many purposes, it is sufficient to enumerate what percentage of the population has earning-power within certain limits; the specification of the individuals does not matter. Statements of this kind are *statistical*, in the sense that they deal with percentages or probabilities but do not specify individuals.   We must therefore study some of the concepts and theorems of probability theory.

4.2   CONCEPTS OF PROBABILITY THEORY—I. PROBABILITY FUNCTIONS

We will study a statistical problem which has nothing to do with physics. Consider a large population of people, several million in number, out of which we select a small sample of 100. Imagine that the height of each of these individuals has been measured. One good way of displaying the information is to plot a *histogram* which we construct as follows. We divide the total possible range of heights into convenient intervals or ranges (say 2 cm wide), choosing them so that there is no ambiguity about classifying an individual whose height is at the end of a range. (For example, if the accuracy of measurement were .01 cm, we might choose ranges 160–161.99 cm, 162–163.99 cm and so on.) Then we count up the number of individuals in each range and plot them as a graph. The result might look like Fig. 4.3(a). The number in each range is drawn as a horizontal line covering the interval. From this diagram we can at once see that the population contains many members around 175 cm tall, and that very short and very tall members are rare.

The histogram is not a smooth curve. Of course it consists of a series of steps because we are dealing with intervals of finite width—but sometimes the steps go up when they might be expected to go down. Intuitively one expects that the distribution of heights should be a smoothly varying function, but it is more or less obvious that the irregularities are present because the number of individuals in any range is small. For example, if the 100 people had by chance included only one more member in the range 166–168 cm and one less in the next lower range, then one of the dips would have disappeared. It will be shown in section 7.7.3 that in many circumstances, if the number of individuals in any class is most probably n, then quite probably the number might lie in the range between $(n+\sqrt{n})$ and $(n-\sqrt{n})$. For example, if the number of individuals might be expected to be 10, then (using different samples from the same population) we would probably find counts anywhere between about 7 and 13. If we expected 100 in any range, then counts between 90 and 110 would be common, and so on. The numbers are said to *fluctuate* between certain limits. Now it is obvious that as n becomes larger, $\sqrt{n}$ becomes larger too, but more slowly; so the fluctuations become proportionally smaller, the larger the numbers are. For example, with 10 members in a range, the expected fluctuation is about $\pm 30\%$; with 100 members it is $\pm 10\%$; with 10^6 members it would be only $\pm 0.1\%$ and so on.

We will in future assume that the sample of the population which we take is very large—so large that the number in any range is itself so large that we can neglect fluctuations. We will then get a smoothly stepped histogram like Fig. 4.3(b), which refers to a sample of 10^6 people.

The number in any range depends of course on the total number in the sample, being proportional to it. For many purposes (for example, comparing samples of different size) this is inconvenient and it is better to refer to some standard number for the total sample. The best choice is 1.

To convert Fig. 4.3(*b*) to this standard, all the ordinates of Fig. 4.3(*b*) have simply to be divided by 10^6. If we take the ordinates of all the steps in Fig. 4.3(*b*) they add up to 10^6; in Fig. 4.3(*c*) they add up to 1. The histogram is said to have been normalized to a total population of 1.

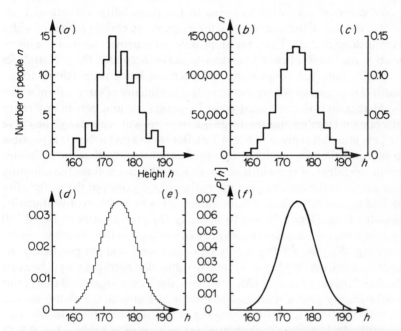

Fig. 4.3. (*a*) Typical histogram showing distribution of heights of 100 people. (*b*) Using scale at left: Expected distribution with 1 million people. (*c*) Using scale at right: Probability of a person's height being within 2 cm range of *h*. (*d*) Using scale at left: Probability of a person's height being within $\frac{1}{2}$ cm range of *h*. (*e*) Using scale at right: Same data, but referring to 1 cm ranges of *h*. (*f*) Probability function *P*[*h*], identical with (*e*) but using infinitesimally small steps.

These ordinates are now called *probabilities*. From Fig. 4.3(*b*) we can read that if we take 10^6 individuals, there would probably be 80,000 whose height lay between 168 and 170 cm. But we can extend this statement by referring to Fig. 4.3(*c*) and saying that if out of the population we selected one single individual at random, the probability that his height lay between these limits would be 0.08. These two statements are equivalent to each other.

This notion of probability is replete with philosophical difficulties which must be faced if a full understanding of many branches of physics is to be reached. But from our present point of view we can regard the probabilities of Fig. 4.3(c) as being merely a convenient shorthand for deriving the *numbers* of Fig. 4.3(b); the number in any range is the probability of being within that range multiplied by the total number in the sample.

We can now ask: what happens to the probability histograms if we reduce the size of the intervals—if for example we choose ranges of width $\frac{1}{2}$ cm instead of 2 cm. First, the steps now get narrower so that the histogram approximates better to a smooth curve. Secondly, the probabilities of lying within each range must now decrease, being proportional to the width of the ranges. (We must emphasize that the size of sample from which the histogram is derived must be so large that the numbers in any range still remain large or else the large fluctuations will reappear.) The curve for $\frac{1}{2}$ cm intervals is given in Fig. 4.3(d). But we can go further. It is possible to plot just the same results in a more useful way, so that the ordinates are independent of the width of the ranges. We can still do the counting and work out the data at $\frac{1}{2}$ cm intervals, but we work out the probability of finding an individual within ranges of *unit width*—conventionally, of width 1 cm. Here, one has to multiply the probabilities of Fig. 4.3(d) by 2, because the standard ranges are twice as wide as the one chosen in the counting. We then get Fig. 4.3(e). Of course we could get practically the same figures by taking Fig. 4.3(c) and dividing *those* ordinates by 2, because the standard range is only half as wide as the 2 cm ranges used there; this would merely give a more crudely stepped histogram. Finally, we can take the limiting case of imagining the intervals to be infinitesimally wide but still calculating the probabilities for unit range of heights, Fig. 4.3(f). The only difference between this and Fig. 4.3(e) is that it is a smooth curve. It is called 'the probability function of h', a phrase which is written $P[h]$.* The equation for the curve of Fig. 4.3(f) is

$$P[h] = 0.067905 \exp[-(h-175)^2/69.031]. \tag{4.1}$$

The value of this function at $h = 185$ is 0.01597; thus the probability that a member of the population has a height between 184.9 and 185.1 cm is $P[h]\,dh = 0.003194$. If there were 10,000 people in a sample, the number of people whose heights lay between these limits would most likely be 32.

The properties of the probability function $P[h]$ may be summarized as follows. The *probability* that an individual has a height within the range dh about h is $P[h]\,dh$. In a population of n people, the *number* with

---

* This symbol must not be thought to imply the same analytical function of the variable every time. For example, we will later meet two probability functions $P[u]$ and $P[c]$ where u and c are respectively a velocity component and the total speed of a molecule; the two are different in analytical form from one another.

heights within these limits is $nP[h]\,dh$. Finally, the fact that the graph has been normalized to a total population of 1 means that by convention

$$\int_0^\infty P[h]\,dh = 1 \qquad (4.2)$$

that is, the area under the graph is unity. In ordinary language, a probability of unity is called certainty, and this equation says that it is certain that the height of any individual lies between zero and infinity, which is correct.

One further point must be mentioned. In many situations where we consider the statistical distributions of the properties of molecules, we can take two quite different points of view. The first is similar to the one we have just considered. We can look at the assembly of molecules *at any one instant* and find the distribution of the relevant quantities (position or momentum coordinates) over them, in just the same way as we did for the heights in a population. But the second procedure has no analogue in the counting of people. It is to follow *a single molecule* for a long time; its coordinates change continually, and we will find that they take on all possible values. For example, the speed of a molecule may be high or low, in any direction, at one time or another. We can then, in principle, find the *fraction of the total time* that the molecule spends in a given state. It is a fundamental assumption that these two kinds of distribution are identical. Thus if 1% of the molecules of a certain liquid have speeds between 10,000 and 10,320 cm/sec (at any one instant), then one single molecule will, for 1% of its time, be travelling with a speed between the same limits. When we come to take averages, we will in other words assume that *sample averages* are identical with *time averages*.

4.2.1   Mean values

The mean value of the height of an individual in a population is defined by

$$\text{mean height } \bar{h} = \frac{\text{total height of all members}}{\text{total number of members}}$$

$$= \frac{\text{sum of terms: (height } h) \times (\text{number with height } h,\, dh)}{\text{total number of members}}$$

that is:

$$\bar{h} = n\int \frac{hP[h]\,dh}{n} = \int hP[h]\,dh \qquad (4.3a)$$

where the integration must be carried out over all possible values of h. For the population of Fig. 4.3, the mean height is

$$\bar{h} = 0.067905 \int_{-\infty}^{\infty} h \exp[-(h-175)^2/69.031] \, dh,$$

where the range of h has to be taken from $-\infty$ to ∞ even though that is unrealistic in practice, the integrand being practically zero over most of the range.

For finding mean values, both in this example and for the functions which will occur later in physical problems, the following integrals will be found useful:

$$\int_{0}^{\infty} e^{-\alpha x^2} \, dx = \frac{1}{2}\sqrt{\frac{\pi}{\alpha}} \qquad \int_{-\infty}^{\infty} e^{-\alpha x^2} \, dx = \sqrt{\frac{\pi}{\alpha}}$$

$$\int_{0}^{\infty} x e^{-\alpha x^2} \, dx = \frac{1}{2\alpha} \qquad \int_{-\infty}^{\infty} x e^{-\alpha x^2} \, dx = 0$$

$$\int_{0}^{\infty} x^2 e^{-\alpha x^2} \, dx = \frac{1}{4}\sqrt{\frac{\pi}{\alpha^3}} \qquad \int_{-\infty}^{\infty} x^2 e^{-\alpha x^2} \, dx = \frac{1}{2}\sqrt{\frac{\pi}{\alpha^3}}$$

$$\int_{0}^{\infty} x^3 e^{-\alpha x^2} \, dx = \frac{1}{2\alpha^2} \qquad \int_{-\infty}^{\infty} x^3 e^{-\alpha x^2} \, dx = 0$$

With these integrals it is not difficult to prove that the mean height in the above example is 175 cm.

In general, the mean value of a quantity x is given by

$$\bar{x} = \int x P[x] \, dx. \tag{4.3b}$$

In the same way we can also find the mean value of any power of x. Later it will be seen that mean square values of certain quantities are significant:

$$\overline{x^2} = \int x^2 P[x] \, dx. \tag{4.3c}$$

This is not in general identical with $\bar{x}^2$; there is no reason why $\int x^2 P[x] \, dx$ should be equal to $(\int x P[x] \, dx)^2$.

4.2.2 Independent probabilities

Consider now a second characteristic of each member of the same population which is quite independent of the first. As an example which we will assume to be independent of height, consider the marks m gained in any particular examination (and let us assume that m is a continuous variable). If h and m were *not* independent, we might find that the examination score was proportional to the height of the examinee or some such relation which we will assume does not hold. We can again measure the probability that any individual scores between m and $(m + \mathrm{d}m)$. Let us write this

$$P[m]\,\mathrm{d}m$$

where $P[m]$ is the probability function of m, perhaps a different function from $P[h]$.

Given these two characteristics h and m, we can ask what is the probability that a given individual has a height between h and $(h+\mathrm{d}h)$ and also scores between m and $(m+\mathrm{d}m)$—a probability which we can write

$$P[h, m]\,\mathrm{d}h\,\mathrm{d}m,$$

where the notation $P[h, m]$ means a function of the two independent variables h and m. We could find this from the data about the population by drawing h and m axes (Fig. 4.4) and representing each individual by a point with his coordinates. Then we could divide up the area into rectangular cells of size $\mathrm{d}h$ by $\mathrm{d}m$, and we could count the number of points inside the appropriate cell; finally we would divide this number by the total population n. In Fig. 4.4, only a few points have been drawn in the cell, but we must assume that the number is really so large that P can be considered as a continuous function of h and m.

We can also relate $P[h, m]$ to $P[h]$ and $P[m]$. For if we choose any person at random from the population the probability that his height is within the desired range is $P[h]\,\mathrm{d}h$. At the same time, the probability that his score is within the desired range is $P[m]\,\mathrm{d}m$. Thus

$$P[h, m]\,\mathrm{d}h\,\mathrm{d}m = P[h]P[m]\,\mathrm{d}h\,\mathrm{d}m. \tag{4.4}$$

This important equation expresses the fact that *for any two characteristics which are entirely independent of one another, the probability of their happening together is the product of their separate probabilities.*

Familiar examples of this occur in games of chance. For example, a die has six faces which are equal except for the markings and the probability of throwing a given number, say a one, is therefore $\frac{1}{6}$. If two dice are thrown, the numbers which turn up are independent of one another, so that the probability of throwing two ones is $(\frac{1}{6})^2$.

Further aspects of probability theory, notably fluctuations, are considered in section 7.7.

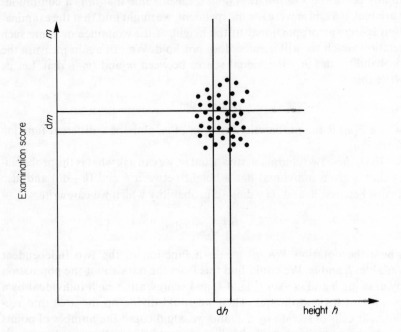

Fig. 4.4. Counting the number of individuals whose height is between h and $(h+dh)$ and score between m and $(m+dm)$ in an examination.

4.3 THERMAL EQUILIBRIUM

We will in what follows consider systems which are in thermal equilibrium; we must try to describe what we mean by this.

From the macroscopic point of view, or from the practical point of view, equilibrium is not too difficult to define even though it is seldom, if ever, strictly attained. The essential conditions are that the temperature must not change with time, however long the system is left, and that the temperature should be uniform throughout the system.

These conditions imply other restrictions. It follows from our definition that there can be no flux or current of heat through the system—for this would produce nonuniformities of temperature. Thus a bar with heat fed into one end and extracted at the other may reach a *steady* state when temperatures do not change with time, but this is not equilibrium because temperatures are not the same everywhere. It follows also from our definition that there cannot be any current or bulk movements of particles through the system, because these would give rise to differences of temperature. Thus a closed vessel partly filled with a liquid is not initially in thermal equilibrium, because the liquid must partly evaporate so that there must be a bulk movement of molecules upwards, driven by a minute temperature difference between liquid and vapour. Only when the space is saturated with vapour and the whole system has settled down to exactly uniform temperature will it be in equilibrium. It follows also from our definition that no system can be in equilibrium when it undergoes variable acceleration, or when dissipative forces are acting within it.

The simple definition of equilibrium is therefore a restrictive one if strictly applied. The atmosphere of the Earth, for example, is not in equilibrium because molecules are escaping from the top of it and because it is absorbing radiation from the sun. Indeed our whole universe is able to function because it is not in equilibrium. But we might take a measurement on some property of a small part of the Earth's atmosphere, and if the changes which took place in it during the time occupied by the measurement were small enough, we could forget that the air was not in strict equilibrium. In a similar way, we can look again at the liquid evaporating into a closed space—strictly speaking, such a system only *approaches* equilibrium asymptotically and never actually *attains* it. But in practice we need often only wait a short time—minutes, say, or days—for the greater part of the change to have taken place. Subsequent changes are likely to be so slow that during the course of any experiment they are negligible. In such cases, equilibrium has for all practical purposes been reached.

From the practical point of view, then, equilibrium is not too difficult an idea to describe. From the molecular point of view, however, all the simplicity of definition evaporates and the situation becomes one of the most elusive to describe. Because of fluctuations, the density in a small volume of gas inside a 'uniform' atmosphere continually changes so that on the atomic scale there are still random fluxes of particles. At the interface of a liquid and a vapour, seemingly quiescent and unchanging on the macroscopic scale, molecules are jumping out and returning all the time. A solid rod is continually expanding and contracting in length, imperfections in the lattice are continually moving about. Let us for the moment then adopt a practical definition of thermal equilibrium and hope that by the time he has reached the end of this book the reader will have a deeper understanding of the meaning of thermal equilibrium at the molecular level.

4.3.1 The Boltzmann law

Let us now return to the problem of specifying the state of an assembly of a large number of molecules. We will assume that the assembly as a whole is at rest, but the molecules have random thermal motion.

In classical mechanics, the state of a molecule must be specified by its coordinates of position, x, y and z, and its components of momentum p_x, p_y, p_z parallel to the three axes. In the ordinary cartesian system, we might as well use velocities instead of momenta (they are related by simple equations of the type $p_x = mv_x$). But since position coordinates and momenta are the conjugate variables used in the Hamiltonian dynamical equations which are valid in all systems of coordinates, we will use momenta here.

Let us now concentrate on one single molecule in an assembly in a container maintained at a temperature T. It does not matter what molecules are being examined, nor what the container is, as long as it functions as a thermostat which is set to maintain the temperature T constant. In fact, we can take an assembly of the same molecules—a portion of a solid embedded in the middle of a larger lump of the solid, for example—considering the whole mass acting as the thermostat for the smaller portion. It is one molecule in this assembly that we concentrate on.

When a system is in thermal equilibrium, the coordinates and momenta of this one molecule are independent of one another. This means that wherever a molecule is located, its momentum may be of any magnitude in any direction. Any interdependence between the coordinates, such as the condition that fast molecules can only be found near the origin, is ruled out. Thus, six numbers or parameters are required to specify the state of one molecule.

For simplicity, let us deal first with p_x, the component of the momentum parallel to the x-axis, of the one molecule whose random thermal motion we are considering.

The energy of this molecule is in general the sum of contributions from all the position coordinates and components of momentum, but the part of the energy contributed by p_x is

$$\mathscr{K}(p_x) = \frac{p_x^2}{2m}, \tag{4.5}$$

where $\mathscr{K}$ stands for kinetic energy and the notation $\mathscr{K}(p_x)$ emphasizes the fact that $\mathscr{K}$ depends upon p_x; m is the mass of the particle and $p_x^2/2m$ might have been written $\frac{1}{2}mv_x^2$ but we have chosen to use the more general formulation.

Then Boltzmann's law states that the *probability* that, in thermal equilibrium, the component of the momentum lies between p_x and $(p_x + dp_x)$ is given by

$$Ae^{-\mathcal{K}(p_x)/kT}dp_x \qquad (4.6a)$$

or in other words the *probability function* for this one component of momentum is

$$Ae^{-\mathcal{K}(p_x)/kT}. \qquad (4.6b)$$

This law is stated at this stage without derivation. The exponential factor itself is called the Boltzmann factor.

In these expressions, A is a quantity which has to be chosen so that the integral of the probability over the whole possible range of p_x is 1—the probabilities have to be normalized.

The quantity T is called the absolute temperature, and these equations define what we mean by absolute temperature. Later we will see that it is identical with the 'perfect gas scale' temperature (section 4.4.1).

The quantity k is a constant, called 'Boltzmann's constant', having the units J/degree, whose magnitude we will work out in section 4.4.1.

This statement of Boltzmann's law is fairly general, but we will now write down an explicit example of a Boltzmann factor and use it to point out a number of features of Boltzmann's law. We choose cartesian coordinates, when the Boltzmann factor for the momentum component p_x is

$$e^{-p_x^2/2mkT}.$$

Let us see what the form of this factor implies—in particular we will show that it is consistent with the intuitive ideas of temperature already described. Firstly, it implies that the most probable value of p_x (the x-component of momentum of the molecule in the assembly in equilibrium) is zero, and that large (positive and negative) values of p_x are less probable. (The distribution follows the same kind of law as the distribution of heights of a population about their mean, the example of section 4.2.) Although for the molecule the mean value of p_x is zero—this is obvious because the distribution is symmetrical about $p_x = 0$, so that positive and negative values of p_x are equally probable—the mean value of p_x^2 is not zero; this in turn means that the contribution to the mean kinetic energy arising from p_x (the mean value of $\mathcal{K}(p_x)$ or $p_x^2/2m$) is not zero. In the next chapter, we work out this mean value and show that it is equal to $\frac{1}{2}kT$, where k is Boltzmann's constant. This is a very important result. It confirms the statement made in section 4.1.2 of this chapter that temperature is a measure of the mean kinetic energy of any molecule in an assembly in equilibrium; we can say therefore that Boltzmann's law is consistent with our intuitive ideas. At the same time it is worth pointing out an elementary mathematical feature of the Boltzmann factor, namely that whereas we usually measure temperature in degrees, (kT) is a measure of the temperature in energy units, ergs or electron volts. The ratio $\mathscr{K}/kT$ is dimensionless—and since one can only take the exponential of a dimensionless number, it is not surprising that the kinetic energy and the temperature are associated in this particular way. Indeed, in Appendix A we demonstrate how the form of the Boltzmann factor can plausibly be derived from simple considerations based on Joule's experiments; we *start* with the fact that the probability function must be a function of (energy)/(temperature) where the temperature is measured in energy units—this is the only mathematically acceptable form it could possibly have.

Having pointed out these aspects of Boltzmann's law, using a simple explicit form of the kinetic energy term, let us return to the more general formulation and extend its use.

The Boltzmann law holds for each of the components of momentum p_x, p_y and p_z independently. For example, the probability that the component of momentum parallel to y lies between the limits p_y and $(p_y + dp_y)$ is

$$A e^{-\mathscr{K}(p_y)/kT} dp_y$$

Combining these two probabilities, we can say that the probability that the x and y components of momentum lie simultaneously between the limits p_x, $(p_x + dp_x)$ and p_y, $(p_y + dp_y)$ is

$$A^2 e^{-\mathscr{K}(p_x)/kT} dp_x \, e^{-\mathscr{K}(p_y)/kT} dp_y.$$

Let us rearrange this expression. When exponentials are multiplied together, we *add* the indices. So we can write the probability

$$A^2 e^{-\{\mathscr{K}(p_x) + \mathscr{K}(p_y)\}/kT} dp_x \, dp_y.$$

By extension, we can calculate the probability that the *three* components of momentum lie simultaneously within limits p_x, $(p_x + dp_x)$; p_y, $(p_y + dp_y)$; p_z, $(p_z + dp_z)$. This is

$$A^3 e^{-\mathscr{K}/kT} dp_x \, dp_y \, dp_z, \tag{4.7}$$

where $\mathscr{K}$ is the total kinetic energy of the molecule:

$$\mathscr{K} \equiv \mathscr{K}(p_x) + \mathscr{K}(p_y) + \mathscr{K}(p_z) = \frac{1}{2m}(p_x^2 + p_y^2 + p_z^2) = \tfrac{1}{2}m(v_x^2 + v_y^2 + v_z^2). \tag{4.8}$$

We might have written $\mathscr{K}(p_x, p_y, p_z)$ to emphasize that the total kinetic energy depends on p_x, p_y and p_z.

The Boltzmann law also holds for the contributions to the energy which come from the position coordinates—these are always potential energies. For example, the potential energy of a molecule undergoing simple harmonic motion is of the form $\mathcal{V}(z) = \frac{1}{2}\alpha z^2$, where z is a displacement and the symbol $\mathcal{V}(z)$ emphasizes the fact that $\mathcal{V}$ depends on z; the potential energy of a molecule in the Earth's gravitational field is of the form $\mathcal{V}(z) = mgz$ where g is the acceleration due to gravity and z is a vertical coordinate. These are two examples of the fact that energies which depend on position coordinates are potential energies.

The Boltzmann law states that the probability that a molecule has a coordinate between z and $(z+dz)$ is

$$(\text{constant})e^{-\mathcal{V}(z)/kT}\,dz,$$

where $\mathcal{V}(z)$ is the contribution to the potential energy which depends on z. The constant is found by normalizing. For a harmonic oscillator, the probability is of the form $(\text{const})\exp(-\alpha z^2/2kT)\,dz$; for a particle in the Earth's gravitational field, it is $(\text{const})\exp(-mgz/kT)\,dz$.

Extending the argument, the probability that the molecule is to be found between $x, (x+dx), y, (y+dy), z, (z+dz)$ is

$$(\text{constant})e^{-\mathcal{V}/kT}\,dx\,dy\,dz, \tag{4.9}$$

where $\mathcal{V}$ is the total potential energy—it might have been written $\mathcal{V}(x, y, z)$ meaning that it depends on x, y and z. Finally, the probability of finding the molecule in the state specified simultaneously by momenta and positions $p_x, (p_x + dp_x) \cdots z, (z+dz)$ is

$$(\text{constant})e^{-E/kT}\,dp_x\,dp_y\,dp_z\,dx\,dy\,dz, \tag{4.10a}$$

where E is the sum of the kinetic and potential energies

$$E = \mathcal{K} + \mathcal{V}$$

and the constant has again to be found by normalizing. Note that in all these expressions, the probability function has always been of the form

$$e^{-E/kT},$$

where E is the energy which depends on the coordinates and momenta considered.

So far we have dealt with the probability of finding a single molecule in a certain state specified by momenta $p_x, (p_x + dp_x), \ldots$ and coordinates $x, (x+dx), \ldots$. It is not difficult to extend the application of Boltzmann's law to two molecules and thence to a large number N.

To do this, we first need a notation. Let us use p_{x1} to denote the x-component of momentum of molecule 1, p_{x2} for the corresponding quantity for molecule 2, and so on. Then we can write down the Boltzmann factors for all the momentum and position coordinates, combine all the exponentials together by adding their indices. The result is that the *probability* of finding one molecule between limits p_{x1}, $(p_{x1} + dp_{x1})$, ..., z_1, $(z_1 + dz_1)$ and simultaneously the second molecule between limits p_{x2}, $(p_{x2} + dp_{x2}) \cdots z_2$, $(z_2 + dz_2)$ is

$$(\text{constant})e^{-E/kT} \, dp_{x1} \cdots dz_1 \, dp_{x2} \cdots dz_2, \qquad (4.10b)$$

where E is now the sum of the kinetic and potential energies of the two molecules. In other words, the *probability function* for this state is

$$(\text{constant})e^{-E/kT}.$$

Similarly, the probability function for N molecules to be found with momentum near $p_{x1}, \ldots, p_{zN}$ and coordinates near $x_1 \ldots z_N$ ($6N$ parameters in all) is

$$(\text{constant})e^{-E/kT},$$

where E is the total energy of the whole assembly. This energy depends on the position and momentum coordinates of all the molecules, and it is (usually) an exceedingly complicated expression. Nevertheless, the *form* of the probability function is so simple that it is consistent with the need to describe a system in equilibrium by only a small number of parameters.

4.3.2 Validity of Boltzmann factors

The problem confronting physicists around the beginning of the century was whether matter really was composed of atoms and molecules, and the statistical methods whose main results we have just presented were evolved in order to tackle this question. To begin with, calculations were performed on exactly specified systems (such as gases, perfect and imperfect) but it was noted that terms of the type $\exp(-E/kT)$ kept appearing. Eventually it was realized that this kind of expression was not a characteristic of any one model of the way a particular lump of matter was constituted but was of the most general validity.

In Appendix A an outline of an approach is given which emphasizes this generality of application. In it, one discusses the thermal equilibrium of a 'subsystem' within a 'system'. The result, that the Boltzmann factor gives the probability function for the subsystem to possess a given energy, is quite independent of what it is made of—provided only that it is in thermal equilibrium. It could be a collection of molecules of the same kind as those of the system, or different from them; it could be a collection of large particles or it could be a single particle while the rest of the system could be a liquid. The subsystem could even be a large object (a metal bar, for example) 'immersed' in a 'gas' of similar but purely imaginary copies of the same object, exchanging energy with one another by some unspecified mechanism whose only function was to ensure that the whole system was in thermal equilibrium. In order to relate the behaviour of the real object in these imaginary surroundings to its behaviour inside a real thermostat we need only assume that the behaviour of any body in thermal equilibrium under given conditions is independent of the mechanism used to bring it to that equilibrium.

The final result of these discussions is that the Boltzmann factor is valid not only for a molecule (which we discuss in section 4.4 and later throughout this book) but for a large object such as a particle undergoing Brownian motion (which we discuss in section 4.4.2).

We will therefore assume that Eq. (4.7) and Eq. (4.9) are both applicable to any object which is in thermal equilibrium and is subject to the laws of classical mechanics. For example, the probability that a particle of microscopic dimensions (large on the molecular scale), suspended in a liquid will be found within the range of coordinates z to $(z + dz)$ is $A \exp(-\mathscr{V}(z)/kT).\,dz$ where $\mathscr{V}(z)$ is the contribution to the potential energy which depends on z.

It is ironic that the intense scrutiny that the laws of statistical mechanics came under resulted in the establishment of their general validity. For this scrutiny was forced on physicists because certain results, notably predictions about specific heats at low temperatures, were in conflict with experiment. But the error lay not in the statistical methods but in the assumption that atoms were subject to classical mechanics.

4.3.3 Kinetic and potential terms

So far we have defined what we mean by a probability function and have quoted the form of the Boltzmann factor. This is the probability function for any system in thermal equilibrium at temperature T to have energy E. In turn, this will allow us to calculate mean values of molecular velocities and other parameters depending on them, and so to calculate the thermodynamic behaviour of many systems

We always begin by writing down an expression for the energy E. For a collection of molecules this is immensely complicated in general, and to reduce it to manageable proportions we have to introduce simplifying assumptions. But in systems obeying the laws of classical mechanics, the Boltzmann factor can always be separated into a product of kinetic and potential energy terms for the molecules. The kinetic energy *always* consists *only* of quadratic terms of the type $p_x^2/2m$ (or $\frac{1}{2}mv_x^2$). Whenever a molecule moves, whether it is moving in a straight line or in an orbit or oscillating about a fixed point, its kinetic energy can always be expressed by terms of this form. Therefore, when we write down a Boltzmann factor for a single molecule or for an assembly, the evaluation of the *kinetic energy term, giving the distribution of velocities, does not depend on the physical nature of the assembly, on whether we are dealing with a solid, liquid or gas.* When we come to calculate velocity distributions or the mean speeds or energies of molecules, this fact leads to some remarkable generalizations.

The potential energy of the whole system is similarly the sum of energies of all the individual molecules. The contribution to this energy arising from fields of force such as gravitational, electromagnetic, electrostatic or magnetic fields, is in general a simple function of the coordinates and the Boltzmann factor breaks up into a product of simple terms. There is a further contribution due to interactions with other molecules, but since the distance apart of each molecule from every other molecule enters into the expression for this potential energy, it is usually an extremely complicated function of the coordinates. It must be emphasized that in contrast to the kinetic energy term, the evaluation of the potential energy term depends very much on the physical nature of the assembly, on whether we are dealing with a solid, liquid or gas.

The plan of the rest of this chapter is as follows. Since the Boltzmann factor incorporates the temperature, and we have stated that this is called the absolute temperature, we must relate it to other scales. We do this by considering (section 4.4) the equilibrium of an assembly of non-interacting particles which is an idealized model of a perfect gas and which bears some resemblance to many real gases. As a result we will be able to identify T and also calculate Boltzmann's constant k. Underlying this discussion will be the need to show that Boltzmann's law gives self-consistent results. Then we will deal with another similar system, a suspension of tiny particles in a liquid (section 4.4.2).

4.4 BOLTZMANN DISTRIBUTIONS—I. A GAS OF INDEPENDENT PARTICLES UNDER GRAVITY

Consider a mass of gas, maintained by some means at a uniform temperature T. It is imagined to be in a very tall vessel—it will emerge later that the results are more interesting when the height is several kilometres—and it is in the gravitational field of the Earth. The problem is to calculate the distribution of the gas molecules with height.

We will not calculate the distribution of kinetic energies (that will be done in the next chapter) but we will deal only with potential energies; the problem can be simplified so that we deal with a single potential energy term and the corresponding Boltzmann factor.

Consider a single molecule. If we assume that the acceleration due to gravity g is constant with height (an approximation which is sufficient for our considerations) then the potential energy of each molecule of mass

m due to its height z above an arbitrary zero (usually at the Earth's surface) is mgz.

Each molecule also has potential energy due to the presence of neighbouring molecules. Most of the time, of course, the molecules are far apart and their interatomic potential energy is negligible. But when they collide with one another, their potential energy is, for a short time, by no means negligible. Now it is essential that there should be collisions between gas molecules because this is the mechanism whereby a gas reaches thermal equilibrium; if it were not for these collisions the gas could never reach uniform temperature. But we will now make a further assumption—namely that the collisions are relatively rare so that *averaged over a long time* the interatomic potential energy is negligible. This can be achieved by ensuring that the gas is at low pressure. It is true that under these conditions, we might have to wait a very long time for the gas to *reach* equilibrium, but that does not concern us. We call such an assembly a *gas of independent particles*.

Thus we reach a compromise: we choose an assembly of molecules where collisions *do* take place, even though the potential energy is large whenever they occur, because collisions are essential to enable thermal equilibrium to be reached. On the other hand, we choose an assembly where the collisions are relatively rare so that *on the average* the interatomic potential energy is negligible. Under these conditions, we can say that the only significant contribution to the potential energy of a molecule is that due to the Earth's gravity, namely mgz, where z is its height above an arbitrary zero.

The probability of finding a molecule at a height between z and $(z+dz)$ is given by the Boltzmann law:

$$P[z]\,dz = Be^{-\mathscr{V}(z)/kT}\,dz = Be^{-mgz/kT}\,dz. \tag{4.11}$$

We can evaluate the constant B by applying the condition that the molecule must be found *somewhere* between the zero of height and the top of the column—which for simplicity we will assume is at $z = \infty$:

$$\int_0^\infty P[z]\,dz = 1 \tag{4.2}$$

whence $B = mg/kT$.

Hence the probability of finding the molecule between z and $(z+dz)$ is

$$P[z]\,dz = \frac{mg}{kT}e^{-mgz/kT}\,dz. \tag{4.12}$$

It is worth remarking that had we chosen any other arbitrary zero for z, all that would have happened is that a constant factor would have appeared in our expression which would later have disappeared in the normalizing.

(Though it does not concern us and the answer is obvious by common sense, we will digress to find the dependence on one of the horizontal coordinates, x. The potential energy does not depend on x: $\mathscr{V}(x) = 0$. Hence the probability of finding the molecule between x and $(x + dx)$ contains the Boltzmann factor $\exp(-0/kT)$ which is unity; and this probability is simply proportional to dx—in other words this implies that the molecule may be found anywhere in the xy plane with equal probability.)

Returning to the z-variation, let us now consider a large number n of molecules. The *number* which we will find between z and $(z + dz)$ is

$$\frac{nmg}{kT} e^{-mgz/kT}\, dz.$$

Now the density of any substance is equal to the number of molecules per unit volume, multiplied by the mass of one molecule. Hence if ρ_0 is the density at zero height (strictly, the limiting density between $z = 0$ and dz where dz is small) then this last equation gives

$$\rho(z) = \rho_0 \exp(-mgz/kT). \tag{4.13}$$

We can plot the variation of density with height, at any given temperature. The density falls off exponentially with height. A graph of $\exp(-z/z_0)$ as a function of z is given in Fig. 4.5; it is identical in form with Fig. 3.2(b). When z is equal to zero it has the value unity. If we compare the function at two heights differing by z_0, say at $z = h$ and $(h + z_0)$, we find that the value falls by a factor $e = 2.717\ldots$; z_0 is called the scale height. At a height of $2z_0$, the function falls by a factor of about 7; at $4z_0$ by a factor of about 50.

Here, the scale height

$$z_0 = \frac{kT}{mg} \tag{4.14}$$

but we cannot yet evaluate this because we do not know the value of Boltzmann's constant k. We will now find this. From this relation between density ρ and temperature T in a gravitational field, we can derive another between ρ and T as a function of pressure P. We do this by changing our approach completely—taking a macroscopic view of the same phenomenon, regarding the gas now as a fluid having weight and capable of exerting a pressure without bothering about its molecular constitution and not

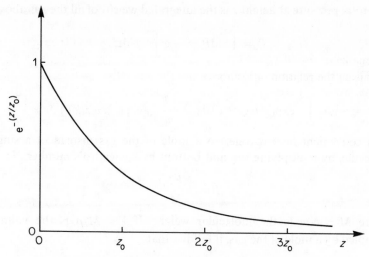

Fig. 4.5. The exponential function $\exp(-z/z_0)$. For an increase of z by an amount z_0, the function decreases by a factor e.

concerning ourselves with *how* the gas exerts its pressure. We consider a tall vessel filled with the gas and deal with the equilibrium of a slice of it between the heights z, and $(z + \mathrm{d}z)$, Fig. 4.6. The pressure at z is greater than that higher up because of the weight of the gas above it; this is true for a column of any material. Thus

$$\mathrm{d}P = -\rho g\,\mathrm{d}z.$$

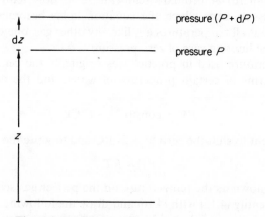

Fig. 4.6. Equilibrium of a gas in the Earth's gravitational field.

The total pressure at height z is the integrated weight of all the gas above:

$$P = \int_z^\infty dP = -g \int_z^\infty \rho \, dz,$$

and using the relation (4.13) above,

$$P = -g\rho_0 \int_z^\infty \exp(-mgz/kT) \, dz = \frac{kT}{m} \rho_0 [\exp(-mgz/kT)]_z^\infty = \frac{kT\rho}{m}.$$

It is convenient here to refer to a mole of the gas instead of a single molecule, by multiplying top and bottom by Avogadro's number N:

$$P = \frac{NkT\rho}{M},$$

where $M = Nm$ is the molecular weight. If $V = M/\rho$ is the volume occupied by a mole of the gas, it follows that:

$$PV = NkT. \tag{4.15}$$

This is a relation between pressure, volume and temperature for the gas of non-interacting particles which we have deduced from Eq. (4.12). It must be remembered that T is still defined by the Boltzmann law (4.11).

4.4.1   Perfect gases

It is well known that real gases obey an equation of just this type, under the correct conditions—namely that the pressure is sufficiently low. Air, for example, obeys this relation with accuracy at pressures of the order of a few atmospheres at ordinary temperature; at low temperatures the range of pressure over which the equation holds becomes very much smaller—but at all temperatures air, like any other gas, obeys this kind of relation in the limiting case of zero pressure.

The temperature used in practice was originally the centigrade scale defined in terms of certain properties of water, and the equation was written

$$PV = \text{const}(273 + T°C).$$

It is convenient to shift the zero to $-273°C$ and to write the equation

$$PV = RT \tag{4.16}$$

where T is known as the temperature on the perfect gas scale.

We can identify (4.15) with (4.16) and draw the following conclusions. Firstly, a perfect gas behaves like a gas of non-interacting particles. Secondly, *the temperature defined by the Boltzmann equation is indeed identical with that on the perfect gas scale*; it was this statement that we set out to prove. Temperature on the absolute scale is written $T\,°K$.

Finally, Boltzmann's constant k can now be found. As a result of measurements on a number of gases at low pressures, extrapolated back to what are called standard conditions of temperature and pressure, namely 1 atmosphere or $1.013 \times 10^6\,dyn/cm^2$, and 0°C or 273°K, it is found that V is then $2.24 \times 10^4\,cm^3$ for all gases and that therefore R is 8.31 J/mol.deg. for all substances.

Since $Nk = R$, and $N = 6 \times 10^{23}$, we have $k = 1.38 \times 10^{-23}\,J\,deg^{-1} = 0.86 \times 10^{-4}\,eV\,deg^{-1}$.

With these data we can now find the scale height of the Earth's atmosphere, making a gross assumption which is quite untrue—that it is all at the same temperature. We use Eq. (4.14). For air, we can take $m = 30\,a.m.u. = 5 \times 10^{-23}\,gm$. Let us take $T = 300°K$ and $g = 10^3\,cm/s^2$. Then $z_0 = kT/mg$ is about 8 kilometres. (We might equally well write $z_0 = RT/Mg$, and take $M = 30$.) Thus if atmospheric pressure is taken to be 76 cm of mercury at the Earth's surface, it is 28 cm at 8 Km a height which is roughly the height of the world's highest mountains. At twice this height the pressure is about 10 cm. (Note that these heights are small compared with the Earth's radius which justifies our approximation that g is constant with height.)

4.4.2 Examples of Boltzmann distributions—II. Brownian movement

Consider a tiny solid particle immersed in a liquid. It exhibits Brownian motion, which means that it can be observed with a microscope to be in ceaseless, rapid motion. It moves randomly about in short discontinuous jumps, though if a higher magnification were used together with a higher speed of observation, each of those jumps would be seen to consist of a number of shorter jumps. These random movements are the result of the impacts of molecules of the liquid against the particle, and this is the mechanism for keeping the temperature of the particle constant on the average.

We can write down the Boltzmann factor for one single particle, and hence calculate a quantity which is almost directly observable: the probability of finding the particle between a height z and $(z+dz)$ from the bottom of the vessel. To do this, we need to know its potential energy at a height z. It is

$$\mathscr{V}(z) = m^*gz$$

where m^* may be called the effective mass of the particle which takes into account the mass of liquid displaced:

$$m^* = v(\rho - \rho')$$

where v is the volume of the particle, ρ and ρ' are the densities of solid and liquid respectively. (This result can be derived from the fact that the force on the particle in the direction of z increasing is $-m^*g$.)

We can therefore write: the probability of the particle being found between the limits of height z and $(z + dz)$, at temperature T, is given by

$$P[z]\, dz = P_0 e^{-m^*gz/kT}\, dz,$$

where P_0 is a normalizing factor which we will not evaluate. If we observed the particle wandering through the liquid over a long time, this expression would give the fraction of the total time it spent within these limits of height. Alternatively, if we had a large number of particles present (large enough for us to apply statistical methods but small enough for us to neglect any mutual attractions between particles), this expression would determine the average number to be found at any time between the given limits:

$$n\, dz = n_0 e^{-m^*gz/kT}\, dz, \tag{4.17}$$

where $n_0\, dz$ is the number between $z = 0$ and dz.

This is just the same form of variation as for molecules of a perfect gas, but now the scale height kT/m^*g is in practice very small instead of being many kilometres because of the magnitude of m^*. Typical numbers are given in a problem at the end of the chapter. The historical significance of this experiment is that it is possible to prepare particles (of a resin, gamboge) whose sizes are great enough for their masses to be found, and yet small enough to undergo Brownian motion with a measurable scale height. Knowing m^* (or more precisely, measuring v, ρ and ρ'), the acceleration due to gravity and the temperature on the perfect gas scale, it was possible firstly to get a direct, almost visual, proof of the validity of the Boltzmann law and secondly (by measuring the scale height) to determine Boltzmann's constant. This was done by Perrin in the early years of this century. Fig. 4.7 is an adaptation of a photograph taken by focusing a microscope at different levels and then making a montage of them; z_0 was of the order of 10^{-3} cm.

4.4.3 Characteristic temperatures

By raising the temperature sufficiently, it is possible to break up any kind of bonding between particles. If at a low temperature the bonding together of the given particles lowers the energy of the assembly by an amount ε per particle, then when there is a reasonably large probability that the energy of thermal agitation is also ε, the interaction will be overcome and the bond broken. The nucleons forming a nucleus, the electrons and nucleus forming an atom, atoms combined together to form a molecule and molecules condensed to form a liquid or solid will all be knocked apart at sufficiently high temperatures.

This will happen above a temperature T given approximately by

$$\varepsilon \approx kT \qquad (4.18)$$

because the Boltzmann factor $\exp(-\varepsilon/kT)$ is then of order unity (we justify this statement a little more precisely in section 5.5). If the temperature were 10 times lower the Boltzmann factor would have the value e^{-10} which is very small, about 10^{-4}, and it would be improbable that sufficient thermal energy would be available.

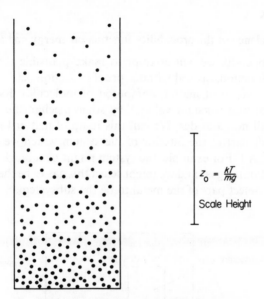

$z_0 = \frac{kT}{mg}$

Scale Height

Fig. 4.7. Distribution of resin particles in water as a function of height. Adapted from a photograph by Perrin (1910), actually a montage of several sections at different heights.

This is a rough rule but it is remarkably powerful. For example, the binding energy of two molecules of many substances is of the order of 10^{-14} to 10^{-13} erg. Correspondingly, each substance has a critical temperature above which the liquid phase can never form, usually of the order of $10^{-14}/k$ to $10^{-13}/k$ degrees—that is, 10^2 to $10^3\,°\mathrm{K}$. Ordinary boiling and melting points are usually not very different in order of magnitude. In a different region of energy, atoms can be ionized at room temperature when they are bombarded with electrons whose energy is of the order of 10 eV (that is, of the order of 10^{-11} erg). We can therefore estimate that the atoms in a gas can be ionized by collisions with other gas atoms at temperatures of the order $10^{-11}/k$ degrees, or roughly $10^5\,°K$. Indeed, gases do form ionized plasmas at such temperatures. Finally, in another and much higher energy range, nuclei can be disrupted by nuclear particles of about 1 MeV energy (around 10^{-6} erg). This means that one can expect nuclear reactions to take place in gases heated to about 10^9 or $10^{10}\,°K$, which corresponds to conditions in the interiors of stars.

APPENDIX A

A.1   Dependence of the probability function on energy and temperature

In this Appendix we will attempt to make plausible the form of the probability function, quoted without proof in section 4.3.1.

Consider a lump of matter, composed of molecules. Its temperature is maintained at a constant value T by some mechanism, a thermostat, which we will not consider. We call this lump of matter the system. We select a small part of the interior of the system which we call the 'subsystem', Fig. A.1. For example, the system might be a block of metal with a cavity containing gas: and we might select this gas to be the subsystem— or we might select part of the metal to be the subsystem.

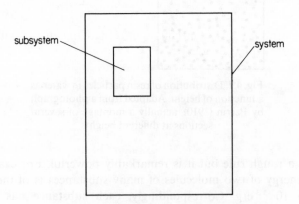

Fig. A.1. A lump of matter (the system) maintained at temperature T; we select part of it, called the subsystem.

We ask what is the probability that the molecules of the subsystem have particular positions and momenta. Since the system is in equilibrium, this probability will not change with time, even though the molecules are moving rapidly and randomly.

If we had just one molecule in the subsystem, we could agree to write

$$P[x, y, z; p_x, p_y, p_z] \, dx \, dy \, dz \, dp_x \, dp_y \, dp_z$$

for the probability that the molecule was at a point whose coordinates lay in the range x to $(x+dx)$, y to $(y+dy)$, and z to $(z+dz)$, while the components of momentum lay between p_x and (p_x+dp_x), p_y and (p_y+dp_y), p_z and (p_z+dp_z).

If we had more molecules—and we usually have large numbers—then we have to write something more complicated. If there were n molecules, we would write

$$P[x_1, y_1, z_1; p_{x_1}, p_{y_1}, p_{z_1}; x_2, y_2, z_2; p_{x_2}, p_{y_2}, p_{z_2}; \ldots x_n, y_n, z_n; p_{x_n}, p_{y_n}, p_{z_n}]$$
$$\times \, dx_1 \, dy_1 \, dz_1 \, dp_{x1} \ldots dp_{y_n} \, dp_{z_n}$$

for the probability that molecule 1 is within a range dx_1 and dy_1 and dz_1 of the point (x_1, y_1, z_1) and moving with a momentum whose components are between limits p_{x_1} and $(p_{x_1}+dp_{x_1})$ and so on; while at the same time molecule 2 is at a point near (x_2, y_2, z_2) with momentum near $(p_{x_2}, p_{y_2}, p_{z_2})$; and so on for all the molecules.

This function P, which is denoted above to be a function of all these variables, could in general be very complicated. The fact that we are dealing with a system in equilibrium however leads to something much simpler.

Since the subsystem is continually interchanging energy with the rest of the system by molecular collisions, it is natural to suppose that P might be a function simply of the energy E of the subsystem. (E itself depends on all the coordinates and momenta of all the molecules.) It also depends on the temperature T which characterizes the system as a whole. No other simple mechanical quantities can be thought of which could be relevant. (The total linear momentum and angular momentum of the system might be included but we rule them out by considering the system as a whole to be at rest.)

Now the probability function is essentially dimensionless because of the meaning of probability. It must therefore be a function of the *ratio* between E and some other quantity which has the dimensions of energy. The only quantity we have at our disposal is T—which is a *measure* of energy, but whose dimensions are largely arbitrary since we can measure temperature in an arbitrary way. The *simplest* possibility is to say that P is a function of (E/kT) where k is a universal constant whose actual magnitude depends upon the units used to measure the temperature:

$$P = f(E/kT). \tag{A.1}$$

This already looks a good deal simpler than our last expression for P, but of course E itself still depends on all the coordinates and momenta and is usually a very complicated expression.

A.2 Form of probability function

We can determine the form of the function f by considering the following special situation. Consider two independent systems both maintained at temperature T; for example, two cavities each containing gas, inside the same block of metal maintained at T. Let the energies of the systems be E_1 and E_2 respectively. The probability of finding system 1 in a given configuration is $f(E_1/kT)$ and the probability of finding system 2 in another given configuration is $f(E_2/kT)$. But if now we consider the two systems together, the probability of finding *both systems* 1 and 2 in their given configurations at the same time must be

$$f\left(\frac{E_1 + E_2}{kT}\right)$$

since the energy of the two systems together is the sum of their separate energies. But since the two systems are independent, the probability must be the product of two separate probabilities:

$$f\left(\frac{E_1 + E_2}{kT}\right) = f\left(\frac{E_1}{kT}\right) f\left(\frac{E_2}{kT}\right). \tag{A.2}$$

We will now show that this serves to determine the form of the function f. This can be seen intuitively by noting that since P (or f) is multiplicative whereas E is additive, the function must be of the kind

$$\log f(E/kT) = BE/kT + C \tag{A.3}$$

where B and C are constants not involving E. Alternatively, this result can be derived explicitly from (A.2) as follows. For simplicity write x_1 and x_2 for E_1/kT and E_2/kT, and write $f'(x)$ for df/dx where x is x_1 or x_2 or $(x_1 + x_2)$. Carry out a partial differentiation of Eq. (A.2) all through with respect to x_1 keeping x_2 constant; then

$$f'(x_1 + x_2) = f'(x_1)f(x_2).$$

Differentiating (A.2) with respect to x_2 keeping x_1 constant gives

$$f'(x_1 + x_2) = f(x_1)f'(x_2).$$

Therefore

$$f'(x_1)f(x_2) = f(x_1)f'(x_2)$$

or

$$\frac{f'(x_1)}{f(x_1)} = \frac{f'(x_2)}{f(x_2)} = B$$

where B is a constant, or more precisely B does not contain E_1 or E_2 although it might depend on T. The solution of this equation is (A.3) above.

Therefore

$$f(E/kT) = (\text{constant})e^{BE/kT}$$

where the constant depends on temperature. Note that we lose no generality by writing B equal to $+1$ or -1 since we can absorb any other numerical factors in the constant k. Thus we have reduced the original complicated form of the expression for the probability of finding the assembly in the specified state to

$$(\text{constant})e^{BE/kT}\, dx_1\, dy_1 \ldots dp_{zn}$$

where B is $+1$ or -1.

Consider now the special case of one molecule in the assembly, which happens to be moving in the x-direction. The energy is simply $p_x^2/2m$, independent of x. The molecule may therefore be found with equal probability to have any x coordinate, but the probability of having momentum between p_x and $(p_x + dp_x)$ is proportional to

$$\exp(Bp_x^2/2mkT)\, dp_x.$$

The top graph in Fig. A.2 shows this function for $B = +1$. It would imply that it is almost certain that the molecule would have infinitely large momentum: this is not acceptable. The lower graph is for $B = -1$, and implies that small momenta are more probable: this is reasonable. Therefore $B = -1$.

Thus we conclude that the probability of finding the assembly in the condition specified is

$$(\text{constant})e^{-E/kT}\, dx_1\, dy_1 \ldots dp_{zn}.$$

The value of the constant of proportionality is determined by the fact that the molecules must be found somewhere inside the accessible range of coordinates and momenta. Thus the integral over all variables must be equal to unity.

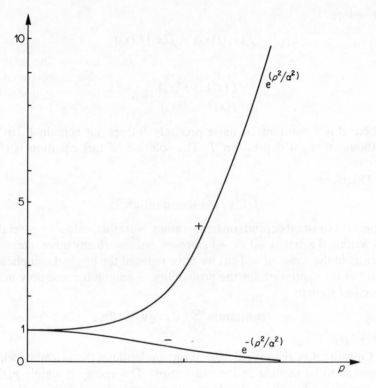

Fig. A.2. Form of the functions $\exp(+p^2/\alpha^2)$ and $\exp(-p^2/\alpha^2)$.

By expressing the energy where possible as the sum of terms each depending on one coordinate or one component of momentum, the probability separates into simple terms each of the form of Eq. (4.6a).

The above expression satisfies all our conditions. It allows the energy of independent systems to be additive but probabilities to be multiplicative. It includes the temperature in a way which agrees with intuitive ideas but at the same time it can serve as the definition of what we mean by temperature and in the text we call it the absolute temperature.

A.3   Extension to macroscopic systems

The language which we used in this discussion of the form of the probability function was a very general one: we spoke of systems and subsystems. Though in the text we concentrated on large assemblies of molecules, and though in deriving the form of the probability function we gave an example of a system consisting of gas molecules inside a cavity, there is no need to limit the discussion in this way. The Boltzmann law can be applied to any system in thermal equilibrium.

For example, consider particles (each containing many molecules) suspended in a liquid and undergoing Brownian motion. The problem (which is discussed in section 4.4.2) is to calculate the distribution of the particles with height, taking into account the potential energy due to the Earth's gravity. We are not now interested in how the individual molecules behave inside each particle but rather in how the particle behaves as a whole; we are interested in how the centre of mass of the whole particle moves, we do not enquire how the molecules move with respect to this centre of mass.

We could begin by accepting the validity of the Boltzmann law for molecules and the distribution of their energies. With this as a starting point, we could then write down the energy of each molecule inside the particle in terms of its own position coordinates and momenta with respect to fixed axes. This expression would be an immensely complicated one. Then we could change the axes of reference to a set through the centre of mass of the particle and moving around with it: this would be a simple linear transformation in which for example the z-coordinate of the nth molecule would be written as $z_n = z'_n + h$ where z'_n is the z-coordinate with respect to the new axes and h the height of the centre of gravity above the fixed origin. The expression for the energy would still be complicated. However, we would be able to group together a number of terms and separate out a term $\exp(-m^*gh/kT)$ from the probability function, where m^* is the effective mass of the *whole* particle. This is the sort of term we are interested in, since both h and m^* refer to the whole particle, or in other words this term does not contain any molecular coordinates or momenta.

But there is no need to go through this complicated procedure. There is no need to assume that the Boltzmann law is valid for molecules only, molecules which are members of large assemblies; hence there is no need to begin each calculation at the molecular level. Our derivation of this law was valid for any system in thermal equilibrium. At one stage for example we considered a system of a block of metal with two cavities each filled with gas; we might as well have considered a liquid containing two particles. A single particle can act as the subsystem and the Boltzmann law applies to it. Thus, we can select coordinates (such as h) which refer to one particle, calculate the corresponding contribution to the energy and hence write down the Boltzmann factor for this particle. This gives exactly the same answer as before, of course, but far more directly.

Two remarks can be made here. First, this discussion was forced upon us because ordinary objects contain vast numbers of molecules and only statistical statements can be made about them. We have however ended by applying the results of this discussion to single objects and this might seem inconsistent. In fact it is not, because the mechanism for keeping its temperature constant is one of continuous bombardment—for example, collisions of a particle with molecules of the liquid in which it is suspended. This situation is so complex that again it cannot be followed in detail.

Secondly, let us go back and examine a little more closely what is involved in accepting that the Boltzmann law is applicable to molecules. After all, each molecule itself has a structure and may contain many, even several hundred particles (electrons, nucleons). Thus if we accepted a strictly classical point of view it would not be sufficient to write down 3 coordinates and 3 momenta for each molecule: many more would be needed. But our discussion has shown that, just as we need not write down the separate Boltzmann factors for every molecule inside a macroscopic particle but can write down the factor for the particle as a whole, so we need not begin by considering the constituent subatomic particles but can deal with the molecule as a whole. At least, then, our approach is self-consistent. (In addition, quantum mechanical considerations show that many types of motion, of the atoms inside a molecule or of the nucleons inside a nucleus which would be expected to occur if classical mechanics were universally valid, do not in fact take place at normal temperatures. The energy needed to excite them is much greater than kT where T is a normally accessible temperature; see section 4.4.3. This merely reinforces the result for completely different reasons.)

PROBLEMS

4.1. In one of his experiments with resin particles suspended in water, Perrin used a microscope with a short depth of focus to count the number of particles in horizontal layers 6 microns apart (1 micron, $\mu = 10^{-6}$ m $= 10^{-4}$ cm). At 17°C the numbers which he observed in the field of view were 305, 530, 940 and 1880. He found the particles to have a radius of 0.52 μ and a density of 1.063 gm cm^{-3}; the density of water at 17°C is 0.999 gm cm^{-3}. Use Perrin's results to calculate Avogadro's number, given that the gas constant R is 8.314 J deg^{-1} mol^{-1} and $g = 980$ cm sec^{-2}.

What would be the distribution with height for particles of density 0.935 gm cm^{-3}?

CHAPTER **5**

The Maxwell speed distribution and the equipartition of energy

5.1 VELOCITY–COMPONENT DISTRIBUTION $P[v_x]$

In Chapter 4, we saw that the probability of an assembly of n molecules possessing coordinates $p_{x1} \cdots z_n$ is given by

$$(\text{const})e^{-E/kT}\, dp_{x1} \cdots dz_n \tag{5.1a}$$

where the x, y and z and the p_x, p_y and p_z coordinates refer respectively to the positions and momenta of the individual molecules. Since the energy splits up into independent contributions, the potential energy $\mathscr{V}$ depending only on the positions and the kinetic energy $\mathscr{K}$ only on the momenta, the probability can be expressed as the product of two entirely independent groups of terms:

$$(\text{const})\, e^{-\mathscr{K}/kT}\, dp_{x1} \cdots dp_{xn} \times e^{-\mathscr{V}/kT}\, dx_1 \cdots dz_n. \tag{5.1b}$$

We then considered an assembly of independent particles in which the potential energy due to the interactions between the molecules was taken to be negligible on the average even though collisions between molecules must occur in order to preserve thermal equilibrium. We did not consider the kinetic energy of the molecules, but by considering a special case of the assembly in the Earth's gravitational field, we showed that the quantity T appearing in the Boltzmann factor is identical with the thermodynamic temperature.

In this chapter we will concentrate on the kinetic energy rather than on the potential and deduce the probability that a molecule is in an assembly in thermal equilibrium at a given temperature and has a momentum component or total momentum or speed, within a certain range.

The results which we will deduce are widely applicable. They hold for any state of matter, solid, liquid or gas, provided the laws of classical physics apply.

To emphasize this point, we will first consider how the total *potential* energy of an assembly of N interacting molecules could be written down—interacting in the sense that each pair of atoms has a mutual potential energy. Considering the interaction of the ith molecule with the jth for example, the potential energy depends on the distance r_{ij} between them raised to the inverse 6th or 12th power (if they are neutral molecules) and to express these quantities in terms of the coordinates of the individual molecules is complicated. To extend this to all possible pairs of molecules gives a vast array of terms which includes many product terms and which cannot be arranged as a simple sum of terms each depending on a single coordinate. This is why in the previous chapter we limited the discussion to one of the few physical cases where there are on the average only negligible interactions—a perfect gas where the molecules are for most of the time at a great distance from one another. Without introducing gross simplifying assumptions, we cannot extend the discussion of the potential energy term to any interacting assembly such as a liquid or a solid or a dense gas.

Even in strongly interacting assemblies, however, the total *kinetic* energy can always be written down as a sum of individual kinetic energies each depending on only one velocity component. For example, in a solid where the molecules are close together and mostly oscillate about their mean positions, the displacement of one molecule from its equilibrium position certainly causes its neighbours to move, either setting them oscillating at a single frequency or with a complicated frequency spectrum, or perhaps changing their places in the lattice. Nevertheless, if the instantaneous velocity of one molecule is (v_{x1}, v_{y1}, v_{z1}) and that of a neighbour is (v_{x2}, v_{y2}, v_{z2}) the energy of each is still the sum of terms of the type $\frac{1}{2}mv_{x1}^2$ or $\frac{1}{2}mv_{x2}^2$, the kinetic energy therefore always separates out into the sum of simple terms and the corresponding Boltzmann factor always separates out into a product of single terms.

Thus the momentum or velocity distribution which we will deduce is applicable to all physical systems. A solid and a liquid and a gas at high pressure and a gas at low pressure, all of the same molecular weight and at the same temperature have the same velocity distribution, even though the types of movements the molecules perform are quite different. In a gas at low pressure the molecule moves in straight lines between relatively rare collisions, and their average interaction potential energy is negligible; in a solid, the molecules are oscillating about their lattice points and have high potential energy. Nevertheless, if we considered a substance of given molecular weight in the gaseous and in the solid states at the same temperature, the proportion of molecules whose velocity lay within a given range would be found to be just the same in the two states. The potential energies would be quite different, the distribution of kinetic energies would be identical. The discussion that follows is in no way limited to a gas—though we shall in fact apply the results to a gas and thereby gain insight into the way the pressure of a gas is produced. Later we shall apply the same results to solids and liquids.

Consider an assembly of molecules each of mass m. It is convenient now to deal with velocities instead of momenta; let v_x, v_y and v_z be the components of velocity parallel to x, y and z, so that $p_x = mv_x$, $p_y = mv_y$, $p_z = mv_z$. Since m is constant, the probability that the x-component of velocity lies between v_x and $(v_x + dv_x)$ is

$$P[v_x]\, dv_x = A \exp(-mv_x^2/2kT)\, dv_x \tag{5.2}$$

where A is determined by the fact that the probability that v_x must have *some* value between $-\infty$ and $+\infty$ is unity:

$$A \int_{-\infty}^{\infty} \exp(-mv_x^2/2kT)\, dv_x = 1.$$

We use one of the integrals listed in the Table on p. 72, namely

$$\int_{-\infty}^{\infty} \exp(-\alpha x^2)\, dx = \sqrt{\frac{\pi}{\alpha}};$$

whence $A = (m/2\pi kT)^{1/2}$ so that the probability distribution is

$$P[v_x]\, dv_x = \left(\frac{m}{2\pi kT}\right)^{1/2} e^{-mv_x^2/2kT}\, dv_x; \tag{5.3}$$

v_x is the x-component of the velocity. A molecule travelling in any direction has a component v_x unless it happens to be moving exactly at right-angles to the x-axis, when v_x is zero. $P[v_y]\, dv_y$ and $P[v_z]\, dv_z$ are identical in form.

A graph of $P[v_x]$ as a function of v_x is given in Fig. 5.1 expressed in dimensionless form. It is a Gaussian distribution centred about the velocity-component $v_x = 0$. The symmetry of the curve means that for every molecule travelling with a certain velocity in the $+x$ direction, it is equally probable that another is travelling in the $-x$ direction. Thus the average value of the total component of momentum is zero, as we would expect in a system at rest.

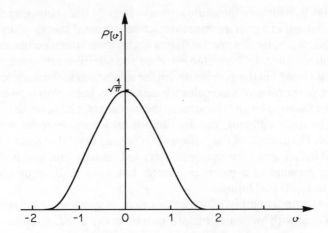

Fig. 5.1. Probability distribution $P[v]$ of a component of molecular velocity in any direction called the x-axis. $v = (mv_x^2/2kT)^{1/2}$, a dimensionless measure of velocity.

5.1.1   Experimental verification of the $P[v_x]$ distribution for gases and solids

It is possible to determine the velocity in any direction of a molecule in thermal equilibrium inside a piece of matter by observing the Doppler shift of a spectral line emitted by the molecule. When stationary, an atom emits a spectral line of a certain wavelength, Λ_0 say, with the same intensity in all directions. An observer can measure this wavelength, the direction of observation being called the x-axis, taken as positive *away* from the observer. When the atom moves, in any direction, the observed wavelength is altered to a value Λ given by

$$\frac{\Lambda - \Lambda_0}{\Lambda_0} = \frac{v_x}{c} \tag{5.4}$$

where v_x is the velocity-component of the atom along the x-axis away from the observer and c is the speed of light. If the atom is moving at any angle to the line of observation, it is always the x-component of the velocity which is measured as a change of wavelength. Radiation can be received from all the atoms in the assembly, whether they are travelling towards the observer and give a shorter wavelength, or receding from him and give a longer wavelength, or moving at right-angles when their wavelength is unaltered. Thus the whole assembly of molecules containing these atoms produces a spread of wavelengths whose intensity at any wavelength is given by the number of molecules with the appropriate value of v_x, that is, by the probability distribution of v_x:

$$I(\Lambda)\,d\Lambda = I(\Lambda_0)\exp[-mc^2(\Lambda-\Lambda_0)^2/2\Lambda_0^2 kT]\,d\Lambda. \tag{5.5}$$

A spectral line which would, in the absence of thermal motion, be sharp is therefore broadened into one of gaussian shape whose width increases with temperature (Fig. 5.2(a)).

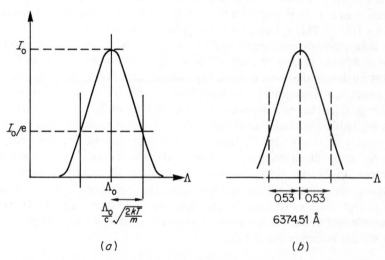

Fig. 5.2. (a) Variation of intensity with wavelength of a spectral line broadened by the Doppler effect due to thermal motion. (b) Broadening of a line emitted by ionized iron atoms ($M = 56$) in the Sun's corona. Data from Dollfus, *Compt. Rend. Acad. Sci.* **236**, 996 (1953).

For observing this effect in gases, and for checking the validity of the $P[v_x]$ distribution curve, optical emission lines in the visible region of wavelengths can be used. Even in gases however, the thermal motion of the molecules is not the only cause of the broadening of spectral lines. Every line has a 'natural width', a quantum effect, determined by the finite lifetime of the excited state of the atom which emits it; many lines however have small natural widths. When molecules collide, the wavelength of the emitted line may be altered by the presence of the nearby molecule, but this effect can be reduced by working at very low pressures where collisions are infrequent. Finally, every optical instrument will broaden a monochromatic line because of its finite resolving power, but this effect can easily be allowed for. Under favourable conditions, therefore, the Doppler broadening can be detected. The agreement with theory is always good.

Interesting applications of this effect have been made in astronomy, to determine the surface temperatures of stars. The Sun's corona, for example, is known to be a very tenuous gas of uniform high temperature. Among the spectral lines it emits is a red line ($\Lambda_0 = 6374$ Å), from highly ionized iron atoms which have lost nine electrons. This line has small natural width, but in the Sun's corona it has been observed to be a comparatively broad line of Gaussian profile, 0.53 Å from the centre of which the intensity falls to $1/e$ of its maximum value Fig. 5.2(b). The molecular weight of iron is 56 and with these data the temperature of the corona is 2.1×10^6 °K. This agrees with other independent estimates.

Solids do not emit sharp visible spectral lines, so that this method of measurement cannot be extended to study the motion of atoms in solids. But nuclear radiations, gamma ray spectra, can under the right circumstances be sharp enough. The complication here is that the radiation is so energetic that the nucleus recoils when it emits a gamma ray (an effect which is negligible with the lower-energy optical radiation) and there is a Doppler broadening due to recoil. However, radiation can be emitted with the recoil momentum being transmitted to the whole crystal instead of being taken up by the single nucleus—a quantum phenomenon called the Mössbauer effect—so that the recoil velocity is negligible, and the emission is sharply monochromatic. In principle, we could then study the Doppler broadening of such a line emitted from a crystal at a finite temperature; it should resemble Fig. 5.2.(a).

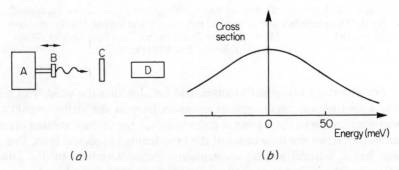

(a)   (b)

Fig. 5.3. (a) Schematic layout of apparatus for measuring the distribution of the velocity-component in solids using the Mössbauer effect. A—device for moving the source rapidly backwards and forwards. B—cooled crystal emitting gamma rays. C—absorbing crystal at room temperature. D—counter which measures gamma rays passing through the absorbing crystal. (b) Absorption cross section per nucleus using iridium emitter and absorber, as a function of energy difference between emitted and absorbed radiation, allowing for recoil energy. The cross section falls by a factor e for 69 meV change of energy. From Visscher, *Ann. Phys.* **9**, 194 (1960).

But there is a further complication because we cannot make a gamma-ray spectrometer to disperse different wavelengths and measure the profile of a line directly to the accuracy required. Instead we make use of the fact that if a given nucleus can decay from an excited state and emit radiation of energy E, then the same nucleus can *absorb* radiation of energy $(E + E_R)$, where E_R is the known energy of recoil, and thereby become excited. (Recoilless absorptions also occur but these can be disregarded.) We utilize this effect as follows (Fig. 5.3(a)). We have two crystals, one a source kept at low temperature so that the emitted gamma rays are monochromatic. It is mounted on a device which oscillates backwards and forwards at high speeds, comparable with thermal speeds of 10^4 or 10^5 cm/s. This motion creates a controllable Doppler shift of frequency of the radiation which then falls on a second crystal. This is the crystal the velocity distribution of whose atoms we wish to explore; it is kept at a high temperature (say, room temperature). We measure the energy absorbed per second as a function of the frequency of the incident gamma rays which we can calculate from the known velocity of the source; we must also allow for the effect of the recoil energy of the absorbing nuclei.

The prediction is that the absorption should show a Gaussian variation with gamma ray energy, of width $2\sqrt{(E_R k T)}$ where T is the temperature of the fixed crystal C and the factor $\sqrt{(kT)}$ comes from the Boltzmann distribution of thermal velocities. For iridium crystals with the absorber at 300°K, the curve is shown in Fig. 5.3(b). (The energy axis has been shifted to allow for the recoil energy and a spike due to recoilless absorptions has been deleted.) For iridium the gamma ray energy is 129 keV, so that $E_R = 0.046$ meV. The essential point is the resemblance of the absorption curve to the Boltzmann curve for velocity-component distribution, Fig. 5.1.

5.1.2   The pressure of a perfect gas

We have repeatedly emphasized that the $P[v_x]$ distribution given by Eq. (5.3) is obeyed by assemblies of molecules in all states of rarefaction or condensation. The first application of this distribution law will, however, be to a perfect gas and we will derive the $PV = RT$ law again.

Consider a gas contained in a vessel, one wall of which is plane, the vessel and gas being in thermal equilibrium. Let the y and z axes be drawn in the plane of the wall, the x axis normal to it. Molecules travel towards the wall, hit it and rebound. Thus each molecule suffers a change of momentum, the force on the wall being the rate of change of momentum. The impacts occur so frequently that this appears as a steady pressure (although refined measurements would show that it does fluctuate about its mean value) which we will now calculate.

We first make the rather unrealistic assumption that each molecule is reflected elastically at impact. In other words, if its initial velocity before impact is (v_x, v_y, v_z) then after impact it is $(-v_x, v_y, v_z)$; the normal component is reversed but the others are unchanged (Fig. 5.4(a)). This implies that the wall must be smooth on the atomic scale, which is of course impossible. At the end of this section we will show that this assumption is unnecessary. Since only molecules travelling towards the wall hit it, v_x may have any value between 0 and ∞ whereas v_y and v_z may have values from $-\infty$ to $+\infty$.

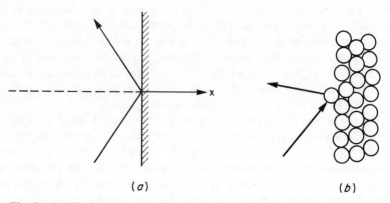

(a) (b)

Fig. 5.4. (a) The impact of a molecule on a wall, assuming an elastic impact and specular reflection from a smooth surface. (b) More realistically, a molecule probably sticks to the wall for a finite time and is reemitted at random.

The nett changes of momentum in the y and z directions from one impact are zero, but in the x direction there is a change of $2mv_x$ per impact.

Let us select out of the whole assembly those molecules with a velocity component between v_x and $(v_x + dv_x)$. If there are a total of n molecules per cm^3, there are $nP[v_x]\,dv_x$ molecules in this class.

Any molecule at a distance equal to, or less than v_x from the wall must hit it during one second. Therefore the number hitting an area A of wall in one second is equal to the number contained in a volume of area A and length v_x.

There are therefore $nAv_xP[v_x]\,dv_x$ impacts per second on area A, each bringing a change of momentum $2mv_x$; so the force on this area is $2mnAv_x^2P[v_x]\,dv_x$ due to these molecules, since the force is the rate of change of momentum. Thus the contribution to the pressure is $2mnv_x^2P[v_x]\,dv_x$.

The total pressure from molecules of all velocities is found by integrating over all relevant values of v_x:

$$P = 2mn \int_0^\infty v_x^2 P[v_x]\, dv_x = 2mn \left(\frac{m}{2\pi kT}\right)^{1/2} \int_0^\infty v_x^2\, e^{-mv_x^2/2kT}\, dv_x.$$

(There should be no confusion between pressure P and probability function $P[\ ]$.) This integral can be evaluated by writing

$$mv_x^2/2kT = \alpha^2, \quad dv_x = \left(\frac{2kT}{m}\right)^{1/2} d\alpha,$$

and by using one of the integrals on page 72:

$$\int_0^\infty \alpha^2\, e^{-\alpha^2}\, d\alpha = \frac{\sqrt{\pi}}{4}.$$

Then $P = nkT$, where n is the number of molecules per cm^3. Let V be the volume of M grams of gas, containing N molecules; then $n = N/V$. Thus

$$PV = NkT = RT.$$

This is the perfect gas law which we have already deduced by considering the potential energy (section 4.4).*

We based this calculation on the assumption that the molecules are reflected specularly on impact, which implies among other things that the walls are smooth compared with molecular dimensions. It is much more realistic to assume that the individual impacts are as shown in Fig. 5.4(b)—a molecule sticks to the wall for a finite time and is reemitted later. In section 9.4.1 we show that 10^{-8} s is a reasonable estimate of the 'sticking time' under certain conditions. While this is a short time in ordinary terms, it is long on the molecular scale; the atoms of the wall perform 1,000 or 10,000 vibrations in that interval. When a molecule jumps off again therefore, it does so at an angle and at a speed unrelated to the incident angle and speed. Let us recast the argument about the momentum change.

Whatever the details of the impacts, we will nevertheless assume that the gas is in equilibrium. Then the number of molecules within any range of velocity must remain constant. Therefore if $nAv_x P[v_x]\, dv_x$ molecules with velocity between v_x and $(v_x + dv_x)$ are removed from the gas in each second by sticking to the area A, the same area must reemit the same number per second to preserve the equilibrium. Therefore, the overall momentum change is the same as we calculated before, even though any one individual molecule may be emitted with quite a different velocity from that at impact. This demonstrates that in order to derive the $PV = RT$ law, we do not

* This cannot pretend to be a new result—it is merely a confirmation of the perfect gas law by a different method.

have to assume any special form of impact at the walls, but merely that the assembly is in thermal equilibrium.

5.2 SPEED DISTRIBUTION $P[c]$

The previous discussion leads to the statement that the probability that any one molecule in any assembly (solid, liquid or gas) in thermal equilibrium has velocity components between v_x and $(v_x + dv_x)$, v_y and $(v_y + dv_y)$, v_z and $(v_z + dv_z)$ is equal to the product of three factors:

$$P[v_x, v_y, v_z] \, dv_x \, dv_y \, dv_z = P[v_x] \, dv_x P[v_y] \, dv_y P[v_z] \, dv_z$$

$$= \left(\frac{m}{2\pi kT}\right)^{3/2} e^{-m(v_x^2 + v_y^2 + v_z^2)/2kT} \, dv_x \, dv_y \, dv_z$$

$$= \left(\frac{m}{2\pi kT}\right)^{3/2} e^{-mc^2/2kT} \, dv_x \, dv_y \, dv_z \qquad (5.6)$$

where we have written the total speed

$$c = (v_x^2 + v_y^2 + v_z^2)^{1/2}. \qquad (5.7)$$

We will now find the probability that the total speed of a molecule lies between limits c and $(c + dc)$. We will no longer be concerned with velocity components v_x, v_y and v_z, or in other words all molecules moving with the same speed will be classed together, no matter which direction they are moving in. We have, in fact, to integrate over all angles. Thus we require an expression of the type $P[c] \, dc$.

5.2.1 Transformation of coordinates

When we were concerned with v_x, v_y and v_z, the velocity of any particle was represented by a point within a framework of (v_x, v_y, v_z) axes, and the number of points within a parallelepipedal element of volume $dv_x \, dv_y \, dv_z$ was counted in a manner analogous to Fig. 4.4. In this new system of counting, the state of the assembly is represented by exactly the same set of points, but the speed c of a molecule is represented by the length of a radius vector from the origin and we no longer have a simply shaped element of volume.

We can transform from $dv_x \, dv_y \, dv_z$ to dc using the following intuitive method. The range between c and $(c + dc)$ is represented by a spherical shell bounded by spheres of radius c and $(c + dc)$, Fig. 5.5. The required number of points, *integrated over all directions*, is therefore proportional to the volume of the shell which is $4\pi c^2 \, dc$. We can therefore write

$$P[c] \, dc = 4\pi \left(\frac{m}{2\pi kT}\right)^{3/2} c^2 \, e^{-mc^2/2kT} \, dc \tag{5.8}$$

for the probability of finding a molecule with speed between c and $(c+dc)$ irrespective of the direction in which it is travelling. This is the Maxwell speed distribution, the result we set out to find.

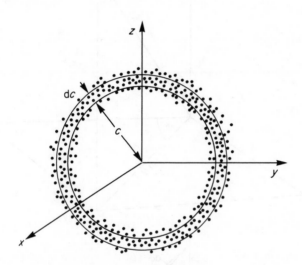

Fig. 5.5. Counting of representative points in a shell
bounded by spheres of radius $c, (c+dc)$.

A more rigorous method of transforming from $dv_x \, dv_y \, dv_z$ to dc is as follows. The velocity of a molecule is uniquely specified by its speed c, the angle θ it makes with an axis (which we may take without loss of generality to be coincident with the z or v_z axis) and another angle ϕ which it makes with a plane through this axis (which we may take to be the xz or $v_x v_z$ plane). This is a system of spherical polar coordinates, Fig. 5.6(a). θ can vary from 0 to π, ϕ from 0 to 2π. When c is varied by dc, θ by $d\theta$, ϕ by $d\phi$, a volume

$$(dc)(c \, d\theta)(c \sin \theta \, d\phi) = c^2 \sin \theta \, dc \, d\theta \, d\phi$$

is generated, Fig. 5.6(b). Thus the probability that the velocity of the molecule lies between c and $(c+dc)$, at angles between θ and $(\theta+d\theta)$, ϕ and $(\phi+d\phi)$ is

$$P[c, \theta, \phi] \, dc \, d\theta \, d\phi = \left(\frac{m}{2\pi kT}\right)^{3/2} c^2 \, e^{-mc^2/2kT} \sin \theta \, dc \, d\theta \, d\phi$$

which is exactly equivalent to Eq. (5.6). $P[c]$ does not contain θ or ϕ, which merely expresses the fact that all directions are equally probable. If this expression is integrated over all possible directions, the little volume element becomes the spherical shell of Fig. 5.5. We then recover the Maxwell distribution (Eq. (5.8)) since

$$\int_0^\pi \int_0^{2\pi} \sin\theta \, d\theta \, d\phi = 4\pi.$$

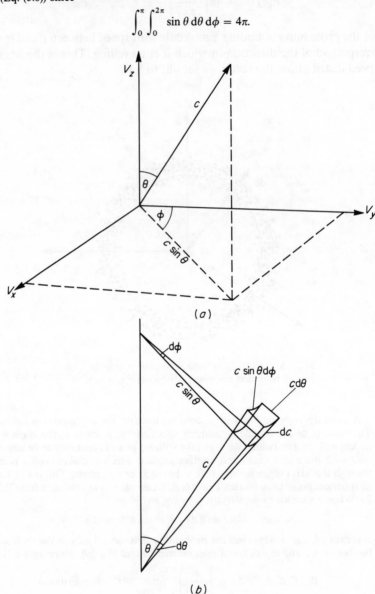

(a)

(b)

Fig. 5.6. (a) Spherical polar coordinates. (b) Generation of volume element by variations dc, $d\theta$, $d\phi$.

5.2.2   The Maxwell distribution

The form of the Maxwell distribution is shown in Figs. 5.7(a) and 5.7(b). The curves are unsymmetrical (in contrast to the curve of Fig. 5.1 for the velocity-component) and pass through the origin and only positive values of speed have any meaning. Writing the ratio

$$\tfrac{1}{2}mc^2/kT = \sigma^2$$

so that σ is a dimensionless quantity proportional to the speed, the distribution law becomes

$$P[\sigma]\, d\sigma = \frac{4}{\sqrt{\pi}}\sigma^2\, e^{-\sigma^2}\, d\sigma$$

and this function is plotted in Fig. 5.7(a). All masses and temperatures are represented by this one graph. The area under the curve is unity—this follows directly from the normalization of the individual components and expresses the fact that the speed of a molecule is certain to lie between zero and infinity. We can graph $P[c]$ directly however, if we select any given mass of molecule and any given temperature. Two such curves, for $M = 28$ and $T = 100°\text{K}$ and $1,000°\text{K}$ respectively are shown in Fig. 5.7(b). The areas under these graphs are also unity. When the temperature is raised, the maximum of the curve moves to higher values of speed, as is to be expected. At the same time, the spread of speeds increases, the curves becoming broader.

Three characteristic values of the speed can be usefully defined—the most probable speed where $P[c]$ goes through a maximum, the mean speed and the root-mean-square speed. They do not differ very greatly from one another.

The most probable speed c_m can be found by setting $dP/dc = 0$ which gives

$$c_m = \left(\frac{2kT}{m}\right)^{1/2}. \tag{5.9}$$

The mean speed $\bar{c}$ is found (following section 4.2.1) from

$$\bar{c} = \int_0^\infty cP[c]\, dc = \frac{2}{\sqrt{\pi}}\left(\frac{2kT}{m}\right)^{1/2} = 1.128\, c_m. \tag{5.10}$$

The mean value of c^2, called the mean square speed $\overline{c^2}$ (which is useful in finding the mean kinetic energy and is not equal to the square of the mean speed), is given by

$$\overline{c^2} = \int_0^\infty c^2 P[c]\, dc = \frac{3kT}{m}. \tag{5.11a}$$

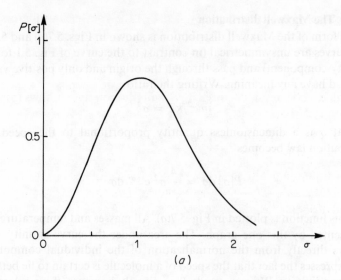

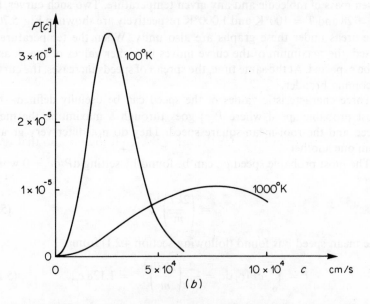

Fig. 5.7. (a) The Maxwell distribution of speeds expressed in terms of the dimensionless measure of speed $\sigma = (mc^2/2kT)^{1/2}$. (b) The Maxwell distribution for $M = 28$, $T = 100°K$ and $T = 1,000°K$. The unit of speed is 1 cm/s. The probability that the speed of a molecule lies between 20,000 and 20,001 cm/s is 3×10^{-5} at 100°K; in 28 g of this substance, there would be nearly 2×10^{19} such molecules.

The square root of this quantity is called the root-mean-square speed c_{rms}:

$$c_{rms} = (\overline{c^2})^{1/2} = \left(\frac{3kT}{m}\right)^{1/2} = 1.225\, c_m. \tag{5.11b}$$

The value of the mean kinetic energy of all the molecules follows from the mean-square speed:

$$\text{mean kinetic energy} = \tfrac{1}{2}m\overline{c^2} = \tfrac{3}{2}kT. \tag{5.12}$$

This result, Eq. (5.12), is a special case of a very general theorem, the equipartition of energy, which will be dealt with at length later in this chapter. In view of its importance, it will be derived again in another way. Instead of finding the mean value of c^2 directly, we may proceed by finding the mean values of v_x^2 and v_y^2 and v_z^2. Since the components are independent variables, the mean values are additive:

$$\overline{c^2} = \overline{v_x^2} + \overline{v_y^2} + \overline{v_z^2}. \tag{5.7}$$

In outline, the calculation is as follows. The mean value of v_x^2 is given by

$$\overline{v_x^2} = \int_{-\infty}^{\infty} v_x^2 P[v_x]\, dv_x = \left(\frac{m}{2\pi kT}\right)^{1/2} \int_{-\infty}^{\infty} v_x^2 \exp(-mv_x^2/2kT)\, dv_x.$$

The same integral has been encountered in section 5.1.2 in the calculation of the pressure of a perfect gas, although the limits of integration are now $-\infty$ to ∞, instead of 0 to ∞. The result is that

$$\overline{v_x^2} = kT/m. \tag{5.13}$$

The same expression holds for the other two contributions, so that once again

$$\overline{c^2} = 3kT/m \tag{5.11a}$$

and the mean kinetic energy is $\tfrac{3}{2}kT$.

5.2.3   Magnitude of the characteristic speeds

The magnitude of these average speeds can be calculated once $(kT/m)^{1/2}$ is known, or $(RT/M)^{1/2}$. Putting $R = 8.31$ J mol^{-1} deg^{-1}, and referring to nitrogen ($M = 28$) and $T = 0°C = 273°K$, this factor is 2.8×10^4 cm/s, so that the r.m.s. speed is nearly 5×10^4 cm/s. Since M varies between 2 and 200 from the lightest to the heaviest element, these speeds all lie in the range 10^4–10^5 cm/s for the elements, whether solids, liquids or gases, at room temperature.

Sound waves consist of ordered movements of molecules, in which energy and momentum are propagated through the medium from molecule to molecule superimposed on the random movements of the molecules. The ordered motion in a solid, for example, might consist of sinusoidal vibrations of the molecules in the direction of propagation or transverse to it, and being a collective mode of motion it can be separated from the random thermal motion on which it is superposed. It is shown in standard texts on wave motion that in all substances the speed of propagation of the sound is comparable with the characteristic speeds of the Maxwell distribution. The calculation for gases is given in section 5.4.2. In air, for example, at 300° K the mean speed is 470 m/s, the speed of sound which consists of longitudinal vibrations only, is 350 m/s. In copper at the same temperature, the mean speed is 316 m/s, while the speeds of longitudinal and transverse sound vibrations through an unbounded volume of the metal are 456 and 225 m/s respectively.

5.2.4   Experimental verification of the $P[c]$ distribution

The Maxwell distribution has been verified experimentally for gases. The most direct methods depend on two techniques—the production of molecular beams and the measurement of their speed distribution using a time-of-flight or chopper method, analogous to Fizeau's method for measuring the speed of light. The experiments of Lammert (1929) are typical. Mercury was heated to 100°C in an oven which had a small hole in one side through which the vapour could escape as a molecular beam. The entire apparatus was under high vacuum. Inside the oven the molecules were practically in equilibrium at 100°C since the rate of loss of molecules was small, so the speeds were distributed according to the Maxwell law. Once a molecule escaped through the hole its speed was not likely to change since it probably never collided with another molecule. Thus though the beam travelled through a space which was not maintained at the same temperature, it was a sample of those molecules inside the oven whose direction of travel happened to lie in the direction of the beam. Let us call this the x-axis—then the method produced molecules whose total velocity vector (of magnitude c) was parallel to x. It was the c-distribution of the beam which was measured; it was not possible to measure the velocity component v_x of molecules not travelling parallel to x, for such molecules were simply not in the beam. Inside the oven, any molecule travelling towards the hole with speed c could escape in time dt if it were within a distance $c\,dt$ of the hole. Hence the number of such molecules escaping per second is proportional to $cP[c]$, which means that the distribution function of the speeds of the molecules in the beam was of the type $c^3 \exp(-mc^2/2kT)$, not the Maxwell distribution though clearly derived from it.

The speeds were found by passing the beam through a velocity selector consisting of two discs (Fig. 5.8(a)), each having 50 narrow radial slits in it, rotating rapidly on a common axis parallel to the beam. The disc further from the oven was turned through a small angle δ with respect to the first. A molecule passing through a slit in the first disc and travelling with a speed c took a time l/c to travel the distance l between discs. If the speed of rotation of the discs on their axis was ω rad/s, the molecule met a slit in the second disc if $\delta = \omega l/c$; or if the angular width of each slit was 2γ, molecules with speeds in the range $\omega l/(\delta + \gamma)$ to $\omega l/(\delta - \gamma)$ could get through both slits; molecules with speeds outside this range were stopped. ω could be varied so as to select different ranges of speeds. In the edge of the discs were wider slits of total angular width equal to that of all the narrow slits together, but these were so wide that they passed molecules

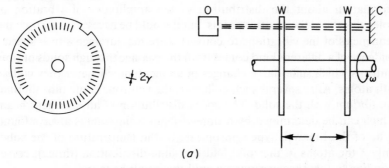

(a)

Fig. 5.8. (a) Slotted wheel with slits acting as a velocity selector. Schematic layout: W—wheels (seen edge on), rotating on common axis; O—oven; C—cold surface. All located inside a high vacuum.

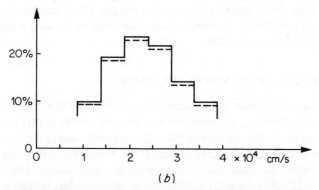

(b)

Fig. 5.8. (b) The percentage of the total intensity of a molecular beam within given limits of speed. Mercury vapour at 100°C. ———— calculated from Maxwell distribution. – – – – – observed. Data from Lammert, Z. Physik **56**, 244 (1929).

of all speeds. The two emergent beams, selected and unselected, fell on a surface cooled with liquid air which trapped the molecules, and their intensities were compared by finding the times needed to produce deposits which were just visible. In one experiment, $l = 6$ cm, $\delta = 4.18°$, $2\gamma = \frac{1}{100}$ rad and the discs rotated at 70 rev/s so that the range between 340 and 390 m/s was selected. The beam through the wide slits was just visible after $4' 40''$, the other after $51' 45''$; thus 9.0% of all the molecules in the beam had speeds in this range. The complete plot is shown in Fig. 5.8(b), which gives the observed intensities and that predicted for a beam with $M = 200$, $T = 373°$K. Other methods of measuring the intensity of the beam have been used.

It has not been possible to measure the speed distribution of molecules in solids directly. X-ray and neutron diffraction measurements can give information about the distribution of the amplitudes of vibration of molecules about their lattice points, but it would be necessary to know the frequencies of the vibrations to convert these measurements into speeds. Evidence of a different kind comes from the passage through solids of slow neutrons which suffer large changes of momentum whenever they collide with atoms. After several such collisions the neutrons come into thermal equilibrium with the solid. The speed distribution of an emergent beam (which can be determined by a time-of-flight technique) is always found to be of the Maxwell type appropriate to the temperature of the solid. Only if the atoms in the solid had the same distribution (though corresponding to their heavier mass) would this result be found.

We will see later that specific heat measurements give an insight into the velocity and speed distributions in solids and that these show that the Maxwell distribution only holds at sufficiently high temperatures. At low temperatures, the motion of the atoms is not described accurately enough by classical mechanics. Quantum mechanics has to be used instead and this leads to different results.

5.3   THE EQUIPARTITION OF ENERGY

When we write down the Boltzmann factor for a particle in an assembly in thermal equilibrium, we have to know its total energy E. So far, we have made use of the fact that E can be split up into kinetic and potential energies which depend on different and independent variables, and by separating those variables we were able to deduce some useful results. One of these was (Eq. (5.12)) that when the kinetic energy of one molecule in an assembly is $\frac{1}{2}mv_x^2 + \frac{1}{2}mv_y^2 + \frac{1}{2}mv_z^2$, then the total kinetic energy of all N molecules in thermal equilibrium at temperature T is $\frac{3}{2}NkT$, and the mean energy of one molecule is $\frac{3}{2}kT$. Thus the mean energy does not depend on v_x or v_y

or v_z in any way but only on the temperature T. It is the purpose of this section to show that this result is a special case of a much more general theorem, which is one of the most important results of classical physics.

It is useful to begin by noting that so far we have not written down explicitly all the forms of energy which a molecule can possess. (We will assume that external fields of force, such as the Earth's gravity, or electric or magnetic fields are absent; this restricts the discussion but the results are nevertheless of significance). The only form of energy which has been explicitly used in calculations has been the translational kinetic energy of the centre of mass, consisting of the three terms of the type $\frac{1}{2}mv_x^2$ mentioned above. We will now consider the energy of angular rotation which every molecule of finite size must possess, and the energy of vibration due to internal oscillations inside a molecule containing more than one atom, or due to the oscillations of a molecule about its equilibrium position inside a solid lattice. Our procedure should really be to consider what conjugate momenta and position coordinates are needed to write down the Hamiltonian expression for the energy. Here we will quote some results without proof and we will use velocities rather than momenta.

First let us consider a rotator, that is a rigid body of arbitrary shape, rotating without any constraint. Then its energy of rotation can be written

$$E_r = \tfrac{1}{2}I_1\omega_1^2 + \tfrac{1}{2}I_2\omega_2^2 + \tfrac{1}{2}I_3\omega_3^2 \tag{5.14}$$

where the I's are the principal moments of inertia and the ω's are the angular velocities about the three mutually perpendicular principal axes of the body. This is the kinetic energy term. We will not consider any potential energy which depends on the angular orientation of the body. Thus, if a molecule of a perfect gas can be considered as a rigid rotator, the energy must have six terms in it, three of the type $\frac{1}{2}mv_x^2$ and three of the type $\frac{1}{2}I\omega^2$. Classically, every ω can vary between $-\infty$ and $+\infty$.

Now let us consider a linear simple-harmonic oscillator, that is a point particle which oscillates about an equilibrium position along a line or, for example, a pair of particles whose oscillation about their centre of mass is in one dimension only. Then two coordinates, a displacement x and velocity v_x are needed to specify the instantaneous state of the oscillator. The energy is partly kinetic, partly potential. For the single particle:

$$E = \tfrac{1}{2}mv_x^2 + \tfrac{1}{2}\alpha x^2$$

where α is the restoring force per unit displacement. It follows that a particle capable of simple harmonic oscillations in three dimensions, such as a molecule in a solid lattice, has six terms in the expression for its energy of oscillation. It is, in principle, possible for all the velocities and all the displacements to take any values between $+\infty$ and $-\infty$. (Notice that E for the one-dimensional oscillator could be written as $\frac{1}{2}\alpha x_0^2$ where x_0 is the amplitude; the sum of the two terms is a constant. But this expression for the energy is not appropriate for the present purpose, since it is required to write it in terms of those coordinates which are needed to specify the *instantaneous state* of the body completely; for an oscillator in thermal equilibrium, the amplitude is not constant. In just the same way, the translational kinetic energy *must* be written in terms of v_x, v_y and v_z and not merely of c.)

Typical terms in these expressions are

$$\tfrac{1}{2}mv_x^2, \qquad \tfrac{1}{2}I\omega^2, \qquad \tfrac{1}{2}\alpha x^2$$

and all of them are of the same type, a constant times the square of a coordinate. Such terms are called degrees of freedom.*

Because of the similarity in the form of these terms we can say at once that angular velocities, for example, are distributed in just the same way as translational velocities. The probability that ω_1, a single component of the angular velocity of a molecule in thermal equilibrium, has a value between ω_1 and $(\omega_1 + d\omega_1)$ is

$$P[\omega_1]\,d\omega_1 = \left(\frac{I_1}{2\pi kT}\right)^{1/2} \exp(-I_1\omega_1^2/2kT)\,d\omega_1$$

an expression completely analogous to $P[v_x]\,dv_x$ (Eq. 5.3). Further it was proved that the mean value of $\frac{1}{2}mv_x^2$ was $\frac{1}{2}kT$ (which follows from Eq. (5.13)). Exactly the same result must hold for the mean value of any of the other terms in the energy. The mean value of $\frac{1}{2}I\omega_1^2$ for an assembly in thermal equilibrium must be $\frac{1}{2}kT$. Exactly the same result must hold for the mean value of $\frac{1}{2}I\omega_2^2$ and $\frac{1}{2}I\omega_3^2$. The mean value of $\frac{1}{2}\alpha x^2$ for an assembly of oscillators in thermal equilibrium must also be $\frac{1}{2}kT$.

This is the theorem of the classical equipartition of energy. Every degree of freedom, that is, every quadratic term in the energy—translational, oscillatory or rotational—contributes $\frac{1}{2}kT$ to the mean energy. An assembly of N particles in thermal equilibrium, each with f degrees of freedom has a mean energy per particle of $\frac{1}{2}fkT$, a total energy of $\frac{1}{2}fNkT$.

The implications may be stated in another way. Consider an assembly of molecules, each one capable of several sorts of motion (oscillation about

---

* It has become customary in statistical mechanics to use the phrase 'number of degrees of freedom' in this way, namely to mean the number of quadratic terms in the energy. The reader should be warned that this is in conflict with the more usual definition in ordinary mechanics which restricts the number of degrees of freedom to the number of *kinetic* energy terms.

a lattice point together with rotation about one or more axes, for example: or translation throughout a volume together with one or more modes of internal oscillation of each molecule). Then according to the equipartition theorem, in thermal equilibrium each possible mode of motion *will* be excited and the amount of energy in each mode is predictable if the temperature is known.

5.4 SPECIFIC HEATS C_p AND C_v

One purpose of the next section is to describe experiments which provide the most direct tests of the validity of the theorem of the equipartition of energy and of the classical mechanics on which it is firmly based. The mean energy of each molecule in a mass of material cannot be measured in any direct experimental way, but specific heats can be measured and these are closely related to it. Our first task will be to define specific heats and to develop some relations between them, then to describe the experimental results. It will emerge that the theorem of equipartition of energy is not of universal validity, and that this is because classical mechanics cannot adequately describe atomic vibrations and oscillations under all conditions and must be replaced by quantum mechanics. Historically the failure of the equipartition theorem to predict the correct specific heats of gases was one of the first symptoms of the inadequacy of classical mechanics ever to be observed.

If dQ is the quantity of heat energy supplied to a standard mass of a substance under certain conditions and the temperature is thereby raised by dT, then the specific heat is defined as dQ/dT under those conditions. The most convenient standard mass is the mole and the appropriate units are J/mol deg. Grams can be used instead of gram mols, and it is occasionally useful to deal with the specific heat per unit volume of a substance.

Practically all bodies when heated under conditions of constant pressure will expand. This expansion absorbs energy in two possible ways. First the external forces acting on the body are pushed back. At the same time, the mean distance between the atoms of the body increases and the potential energy of interaction between atoms is changed—under normal conditions it is increased. It is therefore useful to define two limiting sets of conditions under which specific heats can be measured—constant volume when the body is constrained by external forces not to expand and all the heat energy goes into raising the temperature, and constant pressure where some of the energy is absorbed by the process of expansion. The specific heat at constant volume is denoted by C_v, the specific heat at constant pressure by C_p.

5.4.1   The difference $(C_p - C_v)$

C_v is the quantity which is more readily calculated theoretically for any assembly of atoms or molecules since potential energies of interaction between atoms are constant under conditions of constant volume. In fact if $\bar{E}$ is the mean energy of a standard mass of material, equal to the mean energy of one molecule multiplied by the number of molecules, then

$$C_v = \left(\frac{\partial \bar{E}}{\partial T}\right)_v. \qquad (5.15)$$

C_p is the quantity most easily measured experimentally since laboratory work is usually carried out under conditions of constant pressure. The difference between the two quantities, $(C_p - C_v)$, is therefore of significance, to compare experimental measurements with theoretical predictions.

For a perfect gas, $(C_p - C_v)$ is easy to calculate because the potential energy of interaction between the atoms is always zero. The energy $\bar{E}$, which in thermodynamics is conventionally called the internal energy, therefore does not change when the gas expands at constant temperature. The extra energy absorbed by the expansion at constant pressure must be equivalent only to the work done in overcoming the external pressure. The energy required to produce a given temperature rise under conditions of constant pressure is equal to that needed to produce the same temperature rise when the volume is kept constant, plus the energy absorbed as work done against the external pressure. It has already been shown (4.1.3) that if a body expands in volume by dV against pressure p, the work done is $p\,dV$ (writing the $A\,dx$ of 4.1.3 as dV). Thus

$$C_p\,dT = C_v\,dT + P\,dV$$

or
$$C_p - C_v = P\left(\frac{\partial V}{\partial T}\right)_p \qquad (5.16)$$

It must be stressed that this holds only when the internal energy of the substance does not change with volume. For a perfect gas, $PV = NkT$, whence it follows that

$$C_p - C_v = Nk = R. \qquad (5.17)$$

Different gases can have appreciably different values of the molar specific heats C_p and C_v, but the difference $(C_p - C_v)$ for any one gas must always be equal to 8.31 J/deg, at all temperatures.

For a substance whose internal energy varies with volume, the difference $(C_p - C_v)$ must also vary with volume, and must involve the compressibility and expansion coefficient. It is shown in standard thermodynamic texts that

$$C_p - C_v = \left[P + \left(\frac{\partial \bar{E}}{\partial V} \right)_T \right] \left(\frac{\partial V}{\partial T} \right)_p$$

and using the second law of thermodynamics it may be shown that this can be written

$$C_p - C_v = T \beta^2 K V_0 \tag{5.18}$$

where β is the volume coefficient of thermal expansion, K the isothermal bulk modulus:

$$\beta = \frac{1}{V} \left(\frac{\partial V}{\partial T} \right)_p \tag{5.19}$$

$$K = -V \left(\frac{\partial P}{\partial V} \right)_T \tag{3.10}$$

and V_0 is the molar volume if C_p and C_v are molar specific heats. Inserting typical figures, $(C_p - C_v)$ for most liquids and solids at room temperature are of the order of $R/3$ and $R/10$ respectively. Though the volume expansion is very small, the pressures which would be needed to counteract it are very large so that their product is by no means negligible. For very hard substances such as diamond, $(C_p - C_v)$ is however very small, of the order of $R/1,000$.

Experimental methods for measuring C_p for solids and liquids are straightforward in principle. A known quantity of heat is introduced into the specimen, usually from an electric heater in good thermal contact, and the temperature rise measured, usually with a thermocouple or resistance thermometer. It is essential to reduce heat losses from the specimen as much as possible. The specimen is isolated in a high vacuum and the surroundings are heated separately so as to follow the temperature of the specimen as closely as possible. It is also essential to make sure that the temperature is uniform throughout the specimen before any readings are taken. From the measurements of C_p and a knowledge of the expansion coefficient and compressibility, C_v can be calculated.

5.4.2   Ratio of specific heats C_p/C_v

The methods just described are not very suitable for taking measurements on gases, particularly at the low densities required for them to behave like perfect gases. Their specific heat per unit volume is small, and the heat absorbed by the containing vessel may be comparatively large.

While it is therefore difficult to measure either C_p or C_v directly, it is however simple to measure their ratio

$$\gamma = C_p/C_v. \qquad (5.20)$$

A knowledge of γ, together with the fact that $(C_p - C_v) = R$, allows both C_p and C_v to be found:

$$C_v = \frac{R}{\gamma - 1}, \qquad C_p = \frac{\gamma R}{\gamma - 1}.$$

γ is measured by studying adiabatic changes in a gas.

Adiabatic changes are processes (such as changes of volume) which occur so *slowly* that the assembly of molecules never departs far from thermal equilibrium even though a finite time is required to reestablish the Boltzmann distribution disturbed by the change; and at the same time no heat must flow into the body, $dQ = 0$. The adiabatic compression of a gas must be a *slow* process in which heat cannot escape or enter and since work is done on the gas its temperature must rise. In practice, a compromise has to be struck between the need for slowness of change and the requirement of thermal isolation, and it is usually necessary to accomplish any changes quite quickly.

Sound waves through a gas are adiabatic. They consist of local alternations of compression and rarefaction which produce small local alterations of temperature and these are not dissipated. It is shown in standard texts on wave motion that the speed of a sound wave is given by

$$\text{speed} = \sqrt{\left(\frac{\text{bulk modulus}}{\text{density}}\right)} \qquad (3.21)$$

where the bulk modulus is $-V(\partial P/\partial V)$ under adiabatic conditions. We now calculate this quantity for a perfect gas.

Consider 1 mole of a gas which does work by changing its volume and into which heat also flows, so that the pressure, volume and temperature all change infinitesimally. Then

$$dQ = C_v \, dT + P \, dV \qquad (5.21)$$

Now the relation $PV = RT$ also holds, so that

$$P \, dV + V \, dP = R \, dT$$

and we can use this to eliminate dT:

$$dQ = \frac{C_v}{R}(P \, dV + V \, dP) + P \, dV.$$

Using the relation $C_p - C_v = R$, this becomes

$$R \, dQ = C_p P \, dV + C_v V \, dP.$$

If $dQ = 0$, the adiabatic bulk modulus

$$-V \left(\frac{\partial P}{\partial V} \right)_{ad} = \frac{C_p}{C_v} P = \gamma P, \tag{5.22}$$

so that the speed of sound is given by

$$c_s = \sqrt{\left(\frac{\gamma P}{\rho} \right)} = \sqrt{\left(\frac{\gamma R T}{M} \right)}. \tag{5.23}$$

Measurements of the speed of sound therefore give γ. At the same time, the reason that the speed of sound is comparable with but not equal to the mean speed of the molecules, Eq. (5.10), can be appreciated, since they differ only by a factor $\sqrt{(\pi \gamma / 8)}$ which lies between 0.75 and 0.8 for most gases.

The speed of sound can be measured from the transit time of a pulse of sound (such as a gunshot) between two points a known distance apart, or from simultaneous measurements of frequency and wavelength of a sinusoidal note. Rüchhardt's method is an interesting alternative way of finding the adiabatic bulk modulus γP directly. A resonator consisting of a vessel of a few litres' capacity and filled with the gas under examination is fitted with a vertical glass tube into which a steel ball fits closely, Fig. 5.9. The ball must be accurately spherical, the tube accurately cylindrical and the clearance between the two must be of the order of 0.001 cm so that if the ball is at the top of the tube it sinks only very slowly due to leakage of gas past it. If the ball is given a sudden displacement from its position of near equilibrium it oscillates, typically with a period comparable with one second. The restoring force is the adiabatic elasticity of the gas in the vessel. The system may be considered as a Helmholtz resonator with the ball acting as a heavy driving piston.

If the displacement is x, the change of volume is Ax, where A is the cross sectional area of the tube. Then from the definition of elasticity, the change in pressure is

$$dP = -\gamma P \frac{dV}{V} = -\frac{\gamma P}{V} Ax$$

where V is the volume of the gas. The force on the ball is the pressure times the area

$$F = -\frac{\gamma P A^2}{V} x$$

which shows that the motion is simple harmonic, of period

$$\tau = 2\pi \sqrt{\left/\left(\frac{Vm}{\gamma PA^2}\right)\right.}$$

where m is the mass of the ball. Observation of the period therefore gives γ.

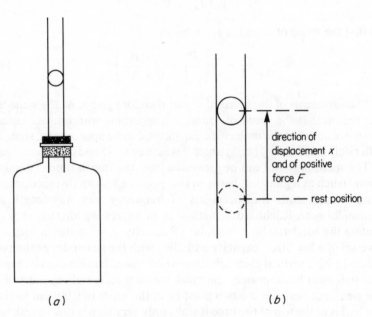

direction of
displacement x
and of positive
force F

—— rest position

(a) (b)

Fig. 5.9. (a) Simplified version of Rüchhardt's apparatus. (b) Ball displaced from rest position.

5.4.3   Results of specific heat measurements

The previous sections have described the principles behind the methods of measuring specific heats and how C_v can be deduced from them. For solids and liquids the most easily measured quantity is C_p and then $(C_p - C_v)$ can be calculated knowing the expansion coefficient and compressibility.

For gases at low pressures the most easily measured quantity is C_p/C_v and since $(C_p - C_v)$ always has the value R, C_v can be calculated. Thus, since $C_v = (\partial \bar{E}/\partial T)_v$, we have a searching method for testing the truth of the law of equipartition of energy.

We will deal with the results for solid elements first. The predicted molar specific heat for a solid having one atom (or ion or rigid molecule) at each lattice point, capable of oscillating harmonically in three dimensions is

$$C_v = N\left(\frac{\partial}{\partial T}\tfrac{6}{2}kT\right)_v = 3R \qquad (5.24)$$

which has a value near 25 J deg^{-1} mol^{-1} at all temperatures. This rule is in fact obeyed by a large number of solid elements at room temperature. It was discovered empirically in the early nineteenth century by Dulong and Petit and was even used as a guide in determining atomic weights in the days when there was some ambiguity about them. But at low temperatures, the specific heat always falls below this predicted value. The graphs for all elements have the same form and can be superposed by merely altering the horizontal scales; Fig. 5.10 shows the curves for three elements. At high temperatures, copper and lead have their expected specific heats of $3R$. At 300°K, however, the value for copper begins to fall off, at 150°K it is decreasing rapidly with temperature, and at 50°K it is very small indeed. Argon does not begin to show a decrease until quite low temperatures are reached. Diamond follows the same sort of curve except that it would only approach its full value of $3R$ if the graph were extrapolated to almost 2,000°K. These results for low temperatures are in sharp conflict with the equipartition law.

The same kind of discrepancy occurs for gases as well. Atoms have finite sizes so that they must each have three principal moments of inertia. An atom of a monatomic gas should therefore have three quadratic terms for its energy of rotation of the type $\tfrac{1}{2}I_1\omega_1^2$ as well as three terms of the type $\tfrac{1}{2}mv_x^2$ for its translational energy. Thus there should be 6 degrees of freedom per atom each contributing $\tfrac{1}{2}kT$ to the energy, so that C_v should be equal to $\tfrac{6}{2}R$ and $\gamma = \tfrac{4}{3} = 1.33$. In fact helium and other rare gases including argon, as well as mercury, all of which are monatomic, are found to have γ quite close to 1.67 so that $C_v = \tfrac{3}{2}R$. Evidently three of the possible degrees of freedom are not excited.

★ ### 5.4.4   Quantum theory and the breakdown of equipartition

These evident failures of the equipartition theorem mean that some of the ideas of classical mechanics itself are at fault. For solids, the explanation of the small specific heats at low temperatures hinges on the fact that the frequency of vibration of the atoms or molecules or ions about their lattice positions is very high, of the order of 10^{12} or 10^{13} cycles per second (see sections 3.6.2, 3.7 and 3.8.2 for estimates of the Einstein frequency). Now it is a result of quantum theory that the energy of an oscillator of frequency v can only take discrete values, $\tfrac{1}{2}hv, \tfrac{3}{2}hv, \tfrac{5}{2}hv$ and so on, where $h = 6.6 \times 10^{-27}$ erg. s or 4×10^{-15} eV. s is Planck's constant. We no longer separate the energy into potential and kinetic contributions, but talk only of the total energy and this can only change by discrete amounts hv.

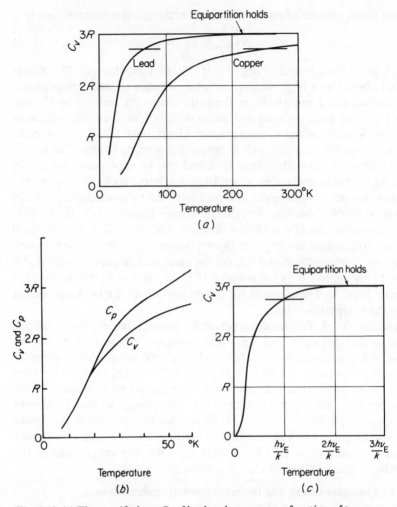

Fig. 5.10. (a) The specific heat C_v of lead and copper as a function of temperature. (b) Specific heat data for solid argon. C_v was derived from the measured C_p using Eq. (5.18); this can be checked using K (the reciprocal of the compressibility) and V_0 from Fig. 3.13 and β derived from the density data:

$T°K$	20	40	60	80
β deg^{-1}	0.4	1.2	1.5	1.8×10^{-3}

(c) Curves for C_v for all substances can be superposed by altering the horizontal scale as shown; the Einstein frequency for copper and lead can then be estimated by comparing the two curves.

In solid organic crystals or solidified rare gases or other soft substances, where v_E is of the order of 10^{12} cycles per second, the energy of a vibrating molecule can only increase in steps of about 7×10^{-15} erg or 0.004 eV. At very low temperatures (when kT is small compared with this quantity, so that the factor $\exp(-hv_E/kT)$ representing the probability of a molecule being in the next higher state is small; to be precise, when T is small compared with 40°K), the oscillators can vibrate only in their lowest state. The energy of oscillation is therefore hardly changed by changing the temperature, so the specific heat is very small. At high temperatures (say 400°K in this case), the discreteness of the vibrational energy levels makes little difference to the mean amount that the oscillators can take up, and the classical law holds. In hard substances like ionic crystals or diamond, where the Einstein frequency is 10 or 50 times greater than in soft crystals, the specific heat is low even at room temperature.

To explain the unexpectedly low specific heats of gases at ordinary temperatures we have to invoke the fact that the energy of a rotating body is quantized. It can only possess discrete values and can only change by discrete amounts. For monatomic helium molecules, the steps are of the order of 10 eV in size; thus the *minimum* energy of rotation that a helium atom may possess is of this order. If its total average energy is very much less than this amount, it is highly improbable that it should be rotating at all. Since kT at room temperature is of the order of $\frac{1}{40}$ eV, it is evident that rotation of the atoms in helium at room temperature does not take place. For monatomic mercury, the steps are of the order of 100 times smaller, but still the thermal motion is not sufficient to excite a significant number of mercury atoms into rotation. Thus they only possess three translational degrees of freedom and the ratio of specific heats is 1.67.

We have just discussed monatomic gases; there is disagreement with the classical equipartition theorem for diatomic molecules also. Many such molecules—like hydrogen, oxygen, chlorine—are dumbell shaped, as in Fig. 5.11. We would expect each molecule to be capable of translation (three energy terms of the type $\frac{1}{2}mv_x^2$ where m is the mass of the molecule

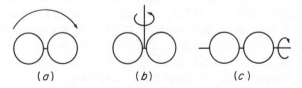

(a)   (b)   (c)

Fig. 5.11. Modes of rotation of a dumbell-shaped diatomic molecule. Specific heat measurements indicate that (a) and (b) occur, (c) does not.

and v_x is the velocity component of its centre of mass), as well as rotation (three terms of the type $\frac{1}{2}I_1\omega_1^2$) and also vibration in and out along the line joining the two atoms as if they were connected by a spring (two terms, for potential and kinetic energies as for any harmonic oscillator). This gives a total of $3+3+2 = 8$ possible degrees of freedom corresponding to $\gamma = \frac{10}{8} = 1.25$. In fact, it is observed that many diatomic gases have γ close to 1.4 (that is, $\frac{7}{5}$) at room temperature which seems to imply that only 5 degrees of freedom are excited at room temperature. Thus, the value for H_2 is 1.408; for N_2 it is 1.405, for NO 1.400 and for O_2 1.396. Again, in order to explain these data we invoke the quantization of energy. The rotation (c) resembles that of a single atom and is again eliminated because the minimum energy of rotation is much larger than kT at ordinary temperatures. Of the 8 degrees of freedom, 7 remain; evidently 2 more must be eliminated. It is not obvious whether these are the two terms in the spring-like internal vibration, or the two rotations (a) and (b) in Fig. 5.11. An exact analysis in fact shows that the vibration is of very high frequency and it is the energy of that motion which is not equi-partitioned; the two rotations (a) and (b) are excited.

The quantization of energy is of course not invoked *only* to explain the specific heat data; other phenomena are also explained at the same time. For gases, the most direct confirmation comes from their spectra, from the interpretation of which it can be confirmed that certain rotations and internal vibrations do not take place at ordinary temperatures.

We have seen that at temperatures where kT is small compared with the energy steps, the value of C_v is lower than predicted on classical equi-partition theory. We would however expect that at sufficiently high tem-peratures, more modes of rotation or vibration will be excited. An increase of C_v and a decrease of γ at high temperatures is therefore to be anticipated. This is indeed found experimentally—notably with hydrogen whose C_v changes from $\frac{3}{2}R$ below 50°K to $\frac{5}{2}R$ at room temperature. This must mean that at 300°K, 5 degrees of freedom are excited but at 50°K only 3. Further, it must mean that two of the modes of rotation or internal vibration must have energy steps whose magnitude is comparable with kT where T is 100°K or 200°K—say 0.01 eV. To take another example, chlorine is ob-served to have γ equal to 1.355 at room temperature and this corresponds neither to 5 degrees of freedom ($\gamma = 1.4$) nor to 6 degrees ($\gamma = 1.33$). Presumably at room temperature 5 of its degrees are fully excited and another is partially excited and the value of γ ought to decrease to 1.33 at temperatures not too far above room temperature.

Among polyatomic molecules, H_2S has $\gamma = 1.340$ which is close to the value $\frac{4}{3}$, corresponding to 6 degrees of freedom; this is taken to mean that the internal vibrations are not excited but all 3 of the rotational modes occur. Larger polyatomic molecules have low values of γ showing that some internal vibrations are excited, presumably because their frequencies are sufficiently low.

Our detailed study of specific heats demonstrates the power as well as the limitations of the equipartition of energy theorem. We may restate this theorem as follows. In an assembly in thermal equilibrium at temperature T, each degree of freedom is excited and contributes $\frac{1}{2}kT$ to the total energy *provided that* kT is large compared with any energy level spacings which an exact quantum-mechanical analysis shows must exist. If the equipartition theorem gives a result which is in conflict with experiment, then this gives some insight into the energy spacings of the modes of motion which are not excited.

5.5 ACTIVATION ENERGIES

We will now use the Maxwell speed distribution to derive a result of general applicability in chemistry as well as physics. It concerns activation energies.

It frequently happens that a system can lose energy provided it can first overcome a barrier. For example, a molecule sitting on the surface of a solid in site X (Fig. 5.12(a)) may be able to find a site Y where its energy is lower. But in order to get there it has to jump or roll over an intervening molecule. Its interatomic potential energy as a function of position therefore resembles that in Fig. 5.12(b). A_0, the height of the barrier which must be jumped, is called the activation energy.

Chemical reactions are often characterized by activation energies. It often happens that two substances can be mixed together without reacting but if they are heated a reaction starts. This reaction may itself give out heat. A well-known example is provided by iron and sulphur which can exist as a mixture at room temperature without undergoing any change, but which when heated react together to form iron sulphide—the reaction, once started, giving out much heat.

The interatomic potential energy as a function of the distance between the centres of two molecules must be of the type in Fig. 5.12(c). As they approach they repel one another (compare Fig. 3.3(a) and (b)) but if they can overcome the barrier of height A_0 then they can stick together. This complex molecule is presumed to initiate or to be the product of the chemical reaction.

We wish to calculate the probability that this takes place. Let us for simplicity assume that the reaction takes place in the gas phase—that is, the two reacting gases F and G are mixed together. Whenever a molecule of F approaches a molecule of G with a kinetic energy greater than A_0, the two can react. Having fallen into the well, the complex molecule must then lose its excess energy but we will assume that this process can take place easily. We call this a reactive collision.

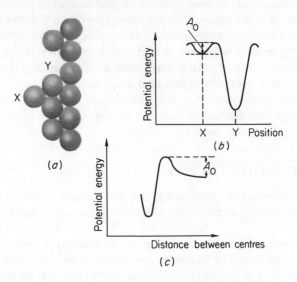

Fig. 5.12. (*a*) and (*b*) A molecule can lose energy by transferring from site X to site Y but needs activation energy A_0 in order to do so. (*c*) Potential energy of two molecules which can undergo a chemical reaction.

The problem of calculating the *relative* speeds of two molecules (and hence the kinetic energy which one imparts to the other on collision) is a complicated one—some of its aspects are dealt with in the next chapter. Let us therefore make two gross simplifying assumptions, firstly that all the molecules are stationary except for one F molecule which is moving with some arbitrary speed c, and further that whenever it meets a G molecule it hits it head on. The effect of these assumptions may (as usual) be expected to make the result incorrect only by a factor of order unity but to leave the form of the result correct—and it is the form which is important here.

Let there be n_G molecules per cm^3 of the species G. In one second, the one moving F molecule travels a distance c. If its area of cross-section is σ (a quantity which is defined more exactly in the next chapter) it sweeps out a volume σc in one second. In this volume there are $n_G \sigma c$ molecules of the other gas; this is therefore the number of collisions it makes in one second with molecules of G (and this result remains true even if the path is not a straight one). The probability that one molecule is indeed moving with speed between c and $(c + dc)$ is given by $P[c] \, dc$, Eq. (5.8), so that if the one F molecule follows the Maxwell distribution it will probably encounter

$$n_G \sigma c P[c] \, dc$$

molecules of G in one second at a speed between c and $(c + dc)$.

If now we imagine n_F molecules of F per cm^3, all of which we can treat in the same way, the number of such encounters is

$$n_F n_G \sigma c P[c] \, dc$$

in each cm^3 per second.

Now let us put in the condition that the kinetic energy $\frac{1}{2}mc^2$ must be equal to A_0 or exceed it, that is $c \geqslant (2A_0/m)^{1/2}$. The number of reactive collisions per unit volume per second is

$$\dot{n}_R = \int_{\sqrt{\frac{2A_0}{m}}}^{\infty} n_F n_G \sigma c P[c] \, dc$$

$$= 4\pi n_F n_G \sigma \left(\frac{m}{2\pi kT}\right)^{3/2} \int_{\sqrt{\frac{2A_0}{m}}}^{\infty} c^3 \, e^{-mc^2/2kT} \, dc$$

after writing $P[c]$ in full, and rearranging.

When realistic modifications are made to allow for the facts that all F and G molecules are moving so that we must calculate the relative kinetic energies, and that not all collisions are head-on, the result happens to be identical with this expression except that a reduced mass $m_F m_G/(m_F + m_G)$ must be used in place of the m.

We can substitute $x = (mc^2/2kT)$ as the variable. This gives

$$dx = \frac{mc}{kT} \, dc$$

$$\dot{n}_R = 2^{3/2} n_F n_G \sigma \left(\frac{kT}{\pi m}\right)^{1/2} \int_{A_0/kT}^{\infty} x \, e^{-x} \, dx.$$

This can be integrated straightforwardly (it is left for the student as an exercise in integration by parts) and the result is a sum of two terms of the form $(kT)^{-1/2} \exp(-A_0/kT)$ and $(kT)^{1/2} \exp(-A_0/kT)$. The important point is that the result contains the factor $\exp(-A_0/kT)$. Because of the rapid variation of $\exp(X)$ with X, the variation of the exponential factor dominates the variation of the number of reactive collisions; over the fairly narrow temperature interval usually encountered, the other factors,

$T^{-1/2}$ and $T^{1/2}$, vary only slowly by comparison and can be regarded as roughly constant. The presence of the factor $\exp(-A_0/kT)$ in the expression for the number of molecules which can jump a barrier of height A_0 is an important result. We have deduced it for a simplified model of a chemical reaction in the gas phase but it is valid for any system where the Maxwell distribution holds. In the system shown in Fig. 5.12(a) and (b) for example, the number of molecules which can jump over the barrier of height A_0 is proportional to $\exp(-A_0/kT)$.

If we can observe experimentally the number of activated molecules (or some macroscopic property which is directly proportional to this number) at different temperatures, then we can deduce the activation energy. The easiest way is to use a graphical method, plotting the number on a logarithmic scale against $1/T$. This gives a straight line whose gradient is $-A_0/k$. When measuring the gradient it is important to remember that an increase of a number by a factor 10 adds 2.303 to its natural logarithm.

The rates of very many chemical reactions have been measured and analysed in this way to find the activation energy. Our simple theory corresponds, in chemists' language, to a bimolecular reaction; they express the rate of chemical change in terms of a 'velocity coefficient' K, which is proportional to the constant connecting $\dot{n}_R$ with the product $n_F n_G$ in our notation. Curves of $\log K$ against $1/T$ and of $\log(\dot{n}_R/n_F n_G)$ against $1/T$ therefore have the same gradient.

To take one example, the reaction

$$CH_3I + C_2H_5ONa = CH_3OC_2H_5 + NaI$$

in alcohol solution was one of the earliest to be measured accurately enough to give consistent results. The concentration of the reactants was determined to begin with, and by measuring the concentration of one of them after the lapse of several minutes the speed of reaction was measured and hence the velocity coefficient obtained. A graph of K on a logarithmic scale against $1/T$ is shown in Fig. 5.13. Note that the reaction rate increases by a factor 40 while the temperature increases only by 10 % from 273°K to 303°K. The slope indicates that the activation energy is 8.1×10^4 J/mol or about 0.85 eV/molecule, an amount comparable with the ionization energy of sodium or iodine.

If we allow ourselves to take a highly simplified view, this fact tells us a good deal about the molecules themselves and the complex which must be formed when the two reacting molecules are in the act of colliding. If the Na in the C_2H_5ONa were in the form of an ion Na^+ and the I were present as I^- in the other reacting molecule, then we would expect an attraction, not a repulsion, between the two molecules. The bonds which link the Na and I atoms inside their respective molecules cannot therefore be ionic but are in fact covalent. During the collision the electrons must be redistributed inside the molecules so as to ionize these atoms—which requires about 1 eV of energy—and sodium iodide can be formed.

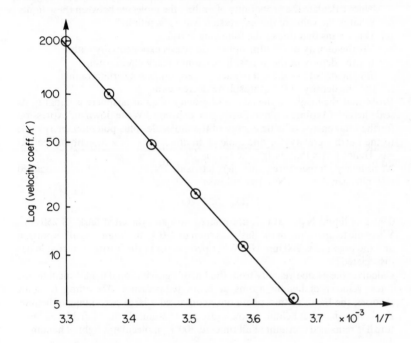

Fig. 5.13. Plot of velocity coefficient K (on logarithmic scale) against $1/T$ for chemical reaction in solution. The temperature varies from 303°K at the left to 273°K at the right. An increase of K by a factor 10 increases $\log_e K$ by 2.303. Data from Hecht and Conrad, Z. Physik. Chem. 3, 450 (1889) reworked in Moelwyn-Hughes Kinetics of Reactions in Solution, Oxford 1947.

PROBLEMS

5.1 The length of the metre is defined in terms of the wavelength of an orange line in the spectrum of the krypton isotope of mass 86 a.m.u., a line which under certain conditions is very narrow. The wavelength is defined to be 6057.8022 Å. With a lamp immersed in liquid nitrogen at 63°K, the width of the line for the intensity to fall by a factor e is 0.0037 Å. (a) Estimate how much of this width is due to Doppler broadening. (b) What would be the width if the lamp were run at 1,000°K?

5.2 (a) A *spherical* planet of mass M has some gas molecules near it. Write down an *exact* expression for the force on a gas molecule at a large distance r from the centre of the planet, and hence for the potential energy. (Use G for Newton's universal constant of gravitation.)

(b) The planet and the molecules are in thermal equilibrium at temperature T. Write down the Boltzmann factor for a molecule at distance r from the centre of the planet. Get the sign of the exponent correct.

(c) Write down the volume of a spherical shell bounded by radii r, $(r+dr)$, and hence calculate the probability of finding the molecule between these limits.

(d) What is the value of this expression when r is infinite?

(e) This means that one of the following is true:
   (i) the density of the atmosphere decreases exponentially with r
   (ii) the density of the atmosphere cannot reach equilibrium
   (iii) the density of the atmosphere is zero outside a certain radius
   (iv) the density of the atmosphere decreases as $1/r$.

5.3 Prove that the height of the centre of gravity of an atmosphere is equal to its scale height. (Assume a plane Earth, g = constant.) Write down an expression for the total energy (kinetic energy of the molecules plus potential energy due to the Earth's gravity) for one mole of an atmosphere of a monatomic perfect gas. Hence prove that its specific heat is $\frac{5}{2}R$.

5.4 At ordinary temperatures, nitrogen tetroxide (N_2O_4) is partially dissociated into nitrogen dioxide (NO_2) as follows:

$$N_2O_4 \rightleftharpoons 2NO_2$$

0.90 g of liquid N_2O_4 at 0°C are poured into an evacuated flask, of 250 cm³. When the temperature in the flask has risen to 270°C, the liquid has all vaporized and the pressure is 960 mm Hg. What percentage of the nitrogen tetroxide has dissociated?

5.5 Calculate the escape velocity from the Earth's gravitational field. Calculate the r.m.s. velocity of helium atoms at room temperature. According to some theories, the Earth's atmosphere once contained a large percentage of helium. Explain the fact that helium is now a rare gas. (Assume that the Earth's temperature has remained constant at all times at 300°K, molecular weight of helium = 4.)

5.6 This problem gives an interesting insight into the energy relations in adiabatic expansions. A piston moves with constant velocity and each molecule undergoes a kind of 'Doppler' change of velocity after reflection from it.

(a) A molecule of mass m approaches a wall with velocity u_x and is specularly reflected. The wall moves with velocity ξ, as shown. Which of the following expressions is correct for the velocity after reflection?

$$-(u_x-2\xi) \qquad -(u_x-\xi) \qquad -(u_x+\xi) \qquad -(u_x+2\xi).$$

Write down the kinetic energy after reflection, assuming that ξ is very small compared with u_x, so that ξ^2 can be neglected.

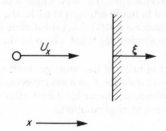

$x \longrightarrow$

(b) Consider a molecule travelling in an arbitrary direction towards the wall, with ζ still in the x-direction. Following an argument like that in section 5.1.2, write down the number of impacts on area A of the wall in a small time $d\tau$, undergone by molecules whose x-component of velocity is between $u_x, (u_x+du)$; hence the change of kinetic energy of this class of molecule in this time. Hence find the change of kinetic energy of all molecules. Express the increase of volume dV in terms of the distance moved by the wall, etc.

(c) Assume the gas is perfect and monatomic. During the small time, the volume is practically constant. Write down the heat capacity of the gas in terms of the specific heat per molecule and the total number of molecules; do not confuse number of molecules/cc with total number of molecules. Hence calculate the change of temperature dT.

(d) Show that $dT/T + 2\,dV/3V = 0$, and that this is consistent with the PV^γ law of adiabatic expansions.

(e) Write down the analogous equation for a polyatomic perfect gas.

(f) What modifications to the discussion are needed if the molecules are not assumed to be reflected specularly?

(g) Describe *briefly* how the ordered motion of the molecules after reflection is degraded into thermal motion (i) in microscopic terms (ii) in macroscopic terms.

5.7 Consider a magnetic dipole in a magnetic field. Using the coordinate system of Fig. 5.6(a) let the vertical axis represent the direction of the field, and the radius vector represent the dipole. From Fig. 5.6(b), it can be seen that for a sphere of unit radius constructed about the origin, an element of area generated by $d\theta$, $d\phi$ is $\sin\theta\,d\theta\,d\phi$.

The potential energy of a dipole of moment m_0 at an angle θ to a field H is $-m_0\mu_0 H\cos\theta$.

For an assembly of N independent dipoles at temperature T :

(i) Write down the probability that a dipole is oriented between $\theta, (\theta+d\theta)$, ϕ, $(\phi+d\phi)$.

(ii) Normalize this expression by integrating over all possible values of θ and ϕ and equating this probability to 1.

(iii) Write down the probability of a dipole being oriented between $\theta, (\theta+d\theta)$, irrespective of the angle ϕ.

(iv) Each dipole has a moment $m_0\cos\theta$ parallel to the field. Write down the contribution from those of N dipoles which are oriented between $\theta, (\theta+d\theta)$.

(v) Hence find the total magnetic moment from all N dipoles. Show that it tends to $Nm_0^2\mu_0 H/3kT$ when $(m_0\mu_0 H/kT)$ is small.
(Note: $\frac{1}{2}(e^x+e^{-x}) = \cosh x$; $\frac{1}{2}(e^x-e^{-x}) = \sinh x$; the limit of $(\coth x - 1/x)$ when x is small is $x/3$.)

5.8 Crystals of sodium chloride show strong absorption of electromagnetic radiation at wavelengths of about 6×10^{-3} cm. Assuming this to be due to the vibrations of individual atoms, calculate (a) the frequency of the vibrations; (b) the potential energy of a sodium atom as a function of its distance δ from its equilibrium position, assuming the vibration to be simple harmonic; (c) the probability distribution of δ at $T = 400°K$; (d) the r.m.s. value of δ at $400°K$. The atomic weight of sodium is 23.

5.9 A galvanometer mirror is suspended on a thread, inside a box containing air. Its moment of inertia for torsional swinging is I. The torsion constant of the fibre is μ. θ is the angle of torsion, ω the corresponding angular velocity. While swinging, the total energy (given by $\frac{1}{2}I\omega^2 + \frac{1}{2}\mu\theta^2$) is constant.

(a) How many degrees of freedom does the system have as regards this motion?
(b) What is the total energy of random swinging of the mirror?
(c) If $\mu = 10^{-6}$ dyn cm, what is the r.m.s. value of the angle of deflection in radians?
(d) Which of the following mechanisms produces these movements?
    (i) expansion and contraction of the mirror, varying I
    (ii) radial expansion and contraction of the fibre, varying μ
    (iii) thermal motion of screw dislocations in the fibre
    (iv) random collisions of air molecules with the fibre, interchanging spin
    (v) random collisions of air molecules with the mirror.
(e) The box is evacuated so that there are very few molecules in it. The box remains at temperature T. Which of the following values does the r.m.s. deflection now have?
    (i) zero
    (ii) reduced in the ratio of the pressures
    (iii) same as before
    (iv) increased in ratio of the pressures
    (v) infinite.
(f) What is the mechanism which produces the movement now?

5.10 The molecules of a substance are known to consist each of two atoms, *rigidly fixed to one another*, like a dumbell. The mass of each atom is 10 a.m.u. so that the molecular weight is 20 a.m.u. Each atom consists of a heavy but extremely small nucleus containing practically all the mass, surrounded by a larger spherical 'cloud' of electrons whose total mass is only about 1/1000th that of the nucleus.
(a) Estimate the order of magnitude of the moment of intertia of the molecule for rotation about the axis shown.
(b) What is the mean kinetic energy in this mode of rotation of one molecule in the solid phase at room temperature (assuming it is not prevented from rotating in any way)?
(c) What is the mean frequency of this rotation at room temperature?
(d) What value of C_v, the specific heat at constant volume, would be expected if the substance were solid at room temperature if the laws of classical physics were applicable?

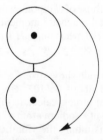

(e) What value of C_v would be expected if the substance were a gas at room temperature if the laws of classical physics were applicable?
(f) What value of $\gamma = C_p/C_v$ would you then expect to find?
(g) What value of γ would you expect to find in practice?

5.11 Hydrogen gas is contained in a thermally insulated cylinder with a moveable piston. If the pressure on the piston is suddenly reduced to 0.38 of its original value as a result the volume of the gas is immediately doubled, estimate the specific heat at constant volume per mole for hydrogen.

5.12 The electrical conductivity σ of a certain class of solid is predicted to vary according to the law

$$\sigma = \frac{C}{T} \cdot e^{-A_0/kT}$$

where C is a constant, $k = 0.86 \times 10^{-4}$ eV/deg and T is the temperature. A_0 is an activation energy, the energy required for an elementary charge to be moved from its atomic site.

Measurements of σ (in arbitrary units) for ice as a function of T are as follows:

σ	31	135	230	630
T	200°K	220°K	230°K	250°K

Plot a suitable straight-line graph and deduce A_0.

Decide whether conduction in ice is dominated by (a) electron conduction with an energy gap $A_0 = 0.1$ eV, (b) proton transport involving the breaking of a hydrogen bond, $A_0 = 0.25$ eV, (c) transport of complex ions requiring simultaneous breaking of 4 hydrogen bonds, $A_0 = 1$ eV.

CHAPTER

6

Transport properties of gases

6.1 TRANSPORT PROCESSES

So far we have concentrated on the properties of solids, liquids and gases which are in equilibrium. In this chapter we will deal with systems which are *nearly but not quite* in equilibrium—in which the density (or the temperature or the average momentum) of the molecules varies from place to place. Under these circumstances there is a tendency for the non-uniformities to die away through the movement—the *transport*—of molecules down the gradient of concentration (or of their mean energy down the temperature gradient or their mean momentum down the velocity gradient). We will define certain *transport coefficients* and show how they can be estimated for gases.

Although the systems we consider are non-uniform in some way and cannot therefore be in thermal equilibrium nor obey the Maxwell speed distribution exactly, we will always make the assumption that the departure from equilibrium is only small. We will therefore assume that no error will be introduced if we take the speed distribution inside any region of the substance to be Maxwellian.

6.1.1 Diffusion as a transport process

Diffusion is the movement of molecules from a region where the concentration is high to one where it is lower, so as to reduce concentration gradients. This process can take place in solids, liquids and gases (though this chapter will be mostly concerned with gases). Diffusion is quite independent of any bulk movements such as winds or convection currents or other kinds of disturbance brought about by differences of density or pressure or temperature (although in practice these often mask effects due to diffusion).

One gas can diffuse through another when both densities are equal. For example, carbon monoxide and nitrogen both have the same molecular weight, 28, so that there is no tendency for one or other gas to rise or fall because of density differences; yet they diffuse through each other. Diffusion can also take place when a layer of the *denser* of two fluids is initially *below* a layer of the lighter so that the diffusion has to take place against gravity. Thus, if a layer of nitrogen is below a layer of hydrogen, a heavy stratum below a light one, then after a time it is possible to detect some hydrogen at the bottom and some nitrogen at the top, and after a very long time both layers will be practically uniform in concentration.

Diffusion coefficients of gas α in gas β can be measured with a suitable geometrical arrangement of two vessels with different initial concentrations together with some method of measuring those concentrations—a chemical method or mass spectroscopy for example. If the rates of change of concentration with time are plotted, the diffusion coefficient can be deduced; the equations describing the process are given in section 6.2.

It is also possible to measure coefficients of self-diffusion, of a gas α in gas α for example. This can be done by using two isotopes having the same shape and size of molecules, the same interaction potential and almost the same mass, but which are nevertheless detectably different—one isotope might be radioactive, the other not. One method is described in section 6.2.1. Mass spectrometer methods can also be used. Applied to a solid, one is described in section 9.6.1.

6.1.2   The diffusion equation

We will begin by taking a macroscopic view of the phenomenon, that is, we will write down equations which involve such variables as concentrations or fluxes but will not specifically mention *individual* molecules. We define the *concentration* of α as the number of molecules n per unit volume, and we consider the simple case where n varies with one coordinate only which we call the x-axis. In Fig. 6.1, the concentration at all points in the plane x is n, at $(x+dx)$ it is $(n+dn)$. Then diffusion takes place down the concentration gradient, from high to low concentrations; we are assuming that bulk disturbances are absent. We next define the *flux* J of particles as the number of particles on average crossing unit area per second in the direction of *increasing* x. Notice that both concentration and flux can be measured in moles instead of numbers of molecules: this is equivalent to dividing all through our equations by Avogadro's number N.

In general, the flux J may change with position x and may also change with time t. In other words, J may be a function of x and t so we write it as $J(x, t)$. Of course, there are circumstances where J may be the same for all x, or where it is constant with time, but the most general situation is that J does depend on both.

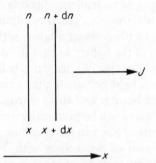

Fig. 6.1. Coordinates used in the definition of diffusion.

It is an experimental fact that, at any instant, the flux at any position x is proportional to the concentration gradient there:

$$J(x, t) \propto -\frac{\partial n}{\partial x}$$

or

$$J(x, t) = -D\frac{\partial n}{\partial x}$$
(6.1)

where D is called the diffusion coefficient. This is known as Fick's law.

By itself, Eq. (6.1) is adequate to describe *'steady-state' conditions* where currents and concentrations do not change with time so that the flux can be written $J(x)$. For example, if a tube of length l cm and constant cross-section A cm^2 has molecules continually introduced at one end and extracted at the other at the same rate, the concentration gradient becomes $-\Delta n/l$, where Δn is the difference of concentration between the two ends. The number of particles crossing any plane in the tube per second is then $-DA\,\Delta n/l$ and this does not change with time.

Consider, however, the much more general situation where initially a certain distribution of concentration is set up and then subsequently the molecules diffuse so as to try to reach a uniform concentration. Concentrations are, therefore, changing with time and particles must be accumulating in the region between x_0 and $(x_0 + dx)$ or moving from it. Therefore, the number crossing area A of the plane x_0 is not equal to that crossing the same area at $(x_0 + dx)$. The flux entering this volume is

$$J_{x_0} = -D\left(\frac{\partial n}{\partial x}\right)_{x=x_0}$$

The flux leaving the slice can be written $J_{x_0 + dx}$ where

$$J_{x_0 + dx} = J_{x_0} + \left(\frac{\partial J}{\partial x}\right)_{x=x_0} dx + \cdots$$

and we can neglect higher terms. The rate of movement of molecules from the slice is equal to the difference between the two values of AJ, and also equal to the volume of the slice, $A\,dx$, times the rate of decrease of n:

$$-\frac{\partial J}{\partial x}A\,dx = \frac{\partial n}{\partial t}A\,dx$$

that is

$$\frac{\partial J}{\partial x} = -\frac{\partial n}{\partial t}. \tag{6.2}$$

Combining this with equation (6.1) and eliminating J:

$$\frac{\partial n}{\partial t} = -\frac{\partial}{\partial x}\left(-D\frac{\partial n}{\partial x}\right) = D\frac{\partial^2 n}{\partial x^2} \tag{6.3}$$

if we assume that D is a constant independent of the concentration. This is called the diffusion equation, and since n depends on x and t it could be written $n(x, t)$.*

Thus we have a system of three equations. (6.1) is an experimental law linking the flux at any point with the concentration gradient there. (6.2) is the continuity equation expressing the fact that molecules cannot disappear, and (6.3) combines these two equations. Eq. (6.1) is adequate for steady-state conditions, where conditions do not vary with time; but for the general case (6.3) may be used.

---

* If the process takes place in 3 dimensions, J is a vector whose components are (J_x, J_y, J_z) and Eqs. (6.1) and (6.2) become

$$\mathbf{J} = \mathbf{i}J_x + \mathbf{j}J_y + \mathbf{k}J_z = -D\left(\mathbf{i}\frac{\partial n}{\partial x} + \mathbf{j}\frac{\partial n}{\partial y} + \mathbf{k}\frac{\partial n}{\partial z}\right) = -D\,\text{grad}\,n$$

$$-\frac{\partial n}{\partial t} = \frac{\partial J_x}{\partial x} + \frac{\partial J_y}{\partial y} + \frac{\partial J_z}{\partial z} = \text{div}\,\mathbf{J}$$

where $\mathbf{i}$, $\mathbf{j}$, and $\mathbf{k}$ are unit vectors parallel to x, y and z. Eliminating J:

$$\frac{\partial n}{\partial t} = -\text{div}(-D\,\text{grad}\,n) = D\nabla^2 n = D\left(\frac{\partial^2 n}{\partial x^2} + \frac{\partial^2 n}{\partial y^2} + \frac{\partial^2 n}{\partial z^2}\right).$$

These are typical of transport equations—with the proviso that for energy and momentum diffusion, the coefficients in the three equations are not all identical as they are here.

6.1.3 Heat conduction

The conduction of heat is also a process of diffusion in which random thermal energy is transferred from a hotter region to a colder one without bulk movement of the molecules themselves. In a hot region of a solid body, the molecules have large amplitudes of vibration; in a hot region of a gas they have extra kinetic energy. By a collision process, this energy is shared with and transferred to neighbouring molecules, so that the heat diffuses through the body though the molecules themselves do not migrate. The macroscopic equations describing conduction in one dimension x are, firstly, the experimental law for the heat flux

$$Q = -\kappa \frac{\partial T}{\partial x} \tag{6.4}$$

(where Q is the heat flux across unit area, measured in $W\,cm^{-2}$, κ is the thermal conductivity and T is the temperature) and, secondly, the continuity equation

$$\frac{\partial Q}{\partial x} = -C\rho \frac{\partial T}{\partial t} \tag{6.5}$$

which expresses the conservation of energy in the form that the heat which is absorbed by a slice of a body goes into raising its temperature. C is the specific heat per unit mass, ρ the density so that $C\rho$ is the specific heat per unit volume. Combining these two equations to eliminate Q:

$$\frac{\partial T}{\partial t} = \left(\frac{\kappa}{C\rho}\right)\frac{\partial^2 T}{\partial x^2} \tag{6.6}$$

where $(\kappa/C\rho)$ is called the thermal diffusivity by analogy with Eq. (6.3). Eq. (6.4) by itself is adequate for steady-state conditions, as when for example heat is fed into one end of a bar and extracted at the other and all temperatures are constant with time, and T can be calculated as a function of x alone. But when conditions are not steady, and T varies with time as well as position, Eq. (6.6) describes the situation.*

6.1.4 Measurement of thermal conductivity of gases

To measure the thermal *diffusivity*, one has to arrange for temperatures to vary with time and to measure the speed of propagation of these

---

* Many students are familiar with (6.4) and with the concept of thermal conductivity but have never met (6.6) and thermal diffusivity. In fact, transient heat flows are of great technical importance.

temperature changes. This is difficult with gases (which we are concentrating on in this chapter), where convection currents may be set up. It is most convenient, therefore, to measure the thermal *conductivity*. Usually this is done by applying Eq. (6.4) directly, using the simplest geometrical arrangement. Commonly, the gas is enclosed between two concentric heavy, metal cylinders. Power is supplied electrically to the inner one; the temperature difference is measured directly with a thermocouple. It is important to correct for the conduction of heat through the electrical leads which can be done by pumping all the gas out and measuring how much heat is still conducted across. It is also important to make sure that the gas does not set up a pattern of convection currents, which it can do rather unexpectedly at certain values of the gap width and pressure. This effect can be detected by using a different size of apparatus and also by checking that the power varies inversely as the gap.

6.1.5   Viscosity

For completeness, a third simple transport process—the diffusion of momentum by viscous forces—will be mentioned here, though only briefly. Viscous motion of fluids can be far more complicated than diffusion or heat conduction and we will be forced to consider only the steady-state equation.

Consider a gas or liquid confined between two parallel plates (Fig. 6.2). Let the lower plate be stationary and the upper plate be moving in the direction shown, which we will call the x-direction. Molecules of fluid very near the plate will be dragged along with it and have a drift velocity, U_x parallel to x, superposed on their thermal velocity. We will assume that U_x is much less than the mean thermal speed or the speed of sound. Molecules of fluid near the stationary plate will, however, remain more or less with zero drift velocity.

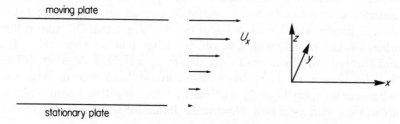

Fig. 6.2. Coordinates used in the definition of viscosity.

Eventually a regime will be set up in which there is a continuous velocity gradient across the fluid from bottom to top. In this state, molecules will be continuously diffusing across the space between the plates and taking their drift momentum with them. Considering an area of a plane parallel to the xy plane in the fluid, molecules which diffuse across from above to below will carry more drift momentum than those which diffuse from underneath to on top. In other words, the more rapidly moving layer tends to drag a more slowly moving layer with it, because of this diffusion of momentum.

In macroscopic terms, a shearing stress (force per unit area) is necessary to maintain this state of motion. The experimental law is

$$P_{xz} = \eta \frac{\partial U_x}{\partial z} \tag{6.7}$$

where P_{xz} is the force per unit area in the x direction due to a gradient of U_x in the z-direction and η is called the coefficient of viscosity. Provided the direction of the force is clearly understood, it is not necessary to include a minus sign, as this depends on the convention for the choice of axes.

We started by considering a fluid in Fig. 6.2, but Eq. (6.7) can be applied to solids because the right-hand side can be written $d\theta/dt$, where θ is an angle of shear. It is difficult to imagine a solid subjected to a shear which goes on increasing with time, but it is quite common for solids to be sheared to and fro in an oscillatory fashion. Forces are then required to provide the accelerations, but in any case the viscosity gives rise to the dissipation of energy and the production of heat. It is usual to refer to this as due to the *internal friction* of solids.

It is implied in Fig. 6.2 that $\partial U_x/\partial z$ is a constant and that U_x increases proportionally to z. This is so if the coefficient η is a constant. For many liquids this holds, but there are notable exceptions where η varies with the velocity gradient or rate of shear so that the velocity profile is not linear. Blood, for example, flows with a much lower viscosity through narrow capillaries than measurements of the flow through wide tubes would indicate—which is fortunate because otherwise one's heart would have to generate several horsepower to maintain circulation. Other suspensions such as cement also have low viscosities when agitated. Oil paint is fluid when worked rapidly with a brush, but when laid on a vertical surface and sheared only by a small force due to its weight, it does not fall off. Such liquids are called thixotropic. Other liquids have opposite behaviour. Whenever we apply Eq. (6.7) to a fluid, therefore, it will be assumed that we are dealing with a gas or a 'Newtonian' liquid for which η is independent of the rate of shear.

When we come to write down equations representing the motion of a fluid while it is not in a steady state but accelerating, we meet a situation which is much more complicated than the diffusion or heat conduction cases. For one thing, there are always mass-acceleration terms which have no analogue in the other phenomena. For another, a kind of regime may be set up where the flow is not streamline as illustrated in Fig. 6.2 but turbulent, and vortices or eddies are present which add an element of randomness to the flow pattern. Whether or not it is set up depends on the ratio of the inertial to the viscous terms. We can, however, usefully adopt a mathematical representation of the simple situation of Fig. 6.2. We can imagine the liquid divided into layers, each one sliding over the one underneath it on imaginary rollers like long axle rods parallel to the y-axis. These rollers are not there in any real sense, but they can lead one to define a quantity called the *vorticity* which is always present in a flowing fluid even when no macroscopic vortices are present. (In a simple case like Fig. 6.2 the vorticity degenerates into the velocity gradient.) Now in the general case of an accelerating fluid with non-uniform velocity it is the vorticity which diffuses throughout the fluid, though the equation it obeys is not of a simple form. For obvious reasons we will not pursue this topic but will be content with the steady-state Eq. (6.7).

6.1.6 Measurement of the viscosity of gases

In his classic experiments to measure the viscosity of gases at low pressures, Maxwell used a torsion apparatus in which a number of circular glass discs were arranged to swing in between fixed ones (Fig. 6.3). He found the *damping coefficient* of the oscillations. If we neglect the energy loss in the torsion wire itself and assume that the discs would go on swinging for a very long time if all the gas were removed, we can calculate the damping as follows.

Consider one surface of one plate, and select an annulus between radii r and $(r + dr)$. Then (assuming streamline flow) the force on this annulus, whose area is $2\pi r \, dr$, is

$$dF = \frac{\eta(r\omega)}{d}(2\pi r \, dr)$$

where the linear velocity is $r\omega$, ω being the angular velocity, and d is the spacing between adjacent moving and stationary surfaces. The contribution to the *couple* is the radius times the force:

$$dG = \frac{2\pi\eta\omega}{d}r^3 \, dr$$

and the total couple is

$$G = \frac{2\pi\eta\omega}{d} \int_0^a r^3 \, dr = \frac{\pi\eta\omega}{2d}a^4$$

where a is the radius of the disc. If there are n discs, each with two surfaces, there are $2n$ such contributions. The equation of motion of the system

when swinging freely is

$$I\frac{d^2\theta}{dt^2}+\frac{(n\pi a^4\eta)}{d}\frac{d\theta}{dt}+\mu\theta = 0$$

where $\omega = d\theta/dt$, I is the moment of inertia, μ the torsion constant of the suspension. This is the equation of a damped oscillation. The time required for the amplitude to decrease by a factor e is $2I/B$, where B is the coefficient of the second term in the equation. Thus η can be determined.

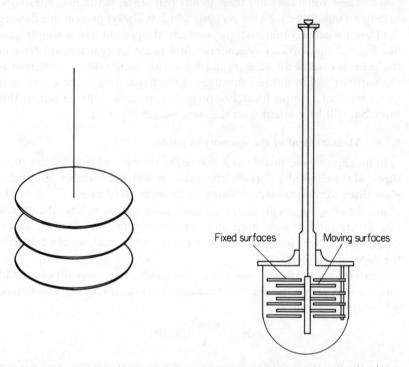

Fixed surfaces   Moving surfaces

Fig. 6.3. Principle of the apparatus for the measurement of viscosity by the damping of torsional oscillations. (a) assembly of discs, (b) section of apparatus.

In Maxwell's final apparatus, there were 3 swinging discs ($n = 3$) with $d = 0.469$ cm. I was determined as 7.33×10^4 g.cm^2; the radius a was effectively 13.1 cm, after allowing for the width of the suspension arrangement in the centre; the period was 72.5 s. In one experiment with air at 21°C, the (natural) logarithmic decrement was determined as 0.073, which meant that 13.7 swings were needed to damp the amplitude by a factor e. From these data, $\eta = 2.47 \times 10^{-4}$ g/cm s. A number of corrections were needed to allow for edge effects and for torsional damping in the suspending wire. This method has been used for measuring the viscosity of liquids as well as gases.

6.2   SOLUTIONS OF THE DIFFUSION EQUATION: THE $\sqrt{t}$ LAW

It is worthwhile studying two solutions of the diffusion equation. The first corresponds to the following initial conditions: A semi-infinite prism of material has area of cross section A; the length is along the x-axis and the ends are at $x = 0$ and $x = \infty$. On the face $x = 0$, N_0 molecules are initially all concentrated in a thin layer and are subsequently allowed to diffuse into the material. We will denote the number at time t which are within a slice between x and $(x + dx)$ by $n(x, t)A\,dx$. Then the appropriate solution of Eq. (6.3) shows that the concentration

$$n(x, t) = \frac{N_0}{A(\pi Dt)^{1/2}}\, e^{-x^2/4Dt} \tag{6.8}$$

The function is shown in Fig. 6.4 for a number of values of the time. The following statements should be verified: (a) that the function does indeed satisfy the diffusion equation, which can be shown by direct substitution, and (b) that the total number of molecules remains constant and equals N_0 at any time t, which can be shown by integrating $n(x, t)A\,dx$ from 0

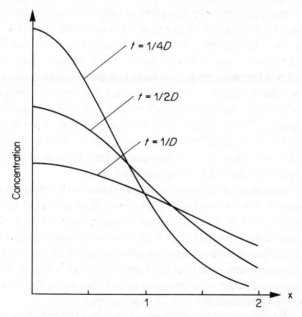

Fig. 6.4. Concentration as a function of x for different values of time.

to ∞, using an integral from the Table on p. 72. It is obvious, from the diagram, that the concentration always remains greatest near the starting place and falls off with increasing distance, and that the spread increases with time, which is all very reasonable.

One very interesting aspect of the diffusion process can be deduced from this solution. On the microscopic scale, diffusion is of course a random process, and it is impossible to predict exactly how far one particular molecule will travel. But if we were to scale down the curves of Fig. 6.4 so as to refer to one molecule instead of N_0, these curves would then be the probability function $P[x]$ for the nett distance travelled by a single molecule at any time (see section 4.2).

We can, therefore, use the curves of Fig. 6.4 to calculate the *mean* nett distance travelled by a molecule at any time t. This is

$$\bar{x}(t) = \frac{A}{N_0} \int_0^\infty x n(x, t) \, dx.$$

Using an integral from the Table on p. 72, we find

$$\bar{x} = \frac{2}{\sqrt{\pi}} (Dt)^{1/2}. \tag{6.9}$$

Thus the mean nett distance travelled is proportional to the *square root* of the time. This is perhaps an unexpected result: one is used to travelling twice as far in twice the time, but for the random process of diffusion this is not so. Of course, some molecules go much further than this, others less far, and it is the *mean* which we have calculated. Stated differently, our result shows that to diffuse a mean distance x, the time required is proportional to x^2. This is an important characteristic of the diffusion process.

Before leaving this problem, note that substituting T for n and taking D to signify the thermal diffusivity, we have the solution to the problem of a semi-infinite slab with a finite amount of heat generated on the surface and subsequently allowed to be conducted away.

Another solution to the diffusion equation refers to the problem of a vessel of cross-section A with two layers (say of liquid or gas, so long as convection is avoided) each of depth $l/2$ and of initial concentration n_0 molecules/cm^3 and zero respectively, Fig. 6.5(a). Diffusion starts at zero time. It is obvious that after an infinite time the concentration throughout the vessel must be uniform and equal to $n_0/2$ molecules/cm^3, Fig. 6.5(b). After a time t, the concentration as a function of distance along the vessel is given by Fig. 6.5(c). (The solution is a Fourier series.)

Initially the concentration difference is n_0, but the mean concentration decreases in one half and increases in the other. The time interval required for the difference of concentration to decrease by a factor e is called the *relaxation time* τ for the diffusion; it is a natural unit of time to use for describing the process. It emerges from the analysis that

$$\tau = l^2/\pi^2 D. \tag{6.10}$$

Note that once again, a time is porportional to the square of a length.

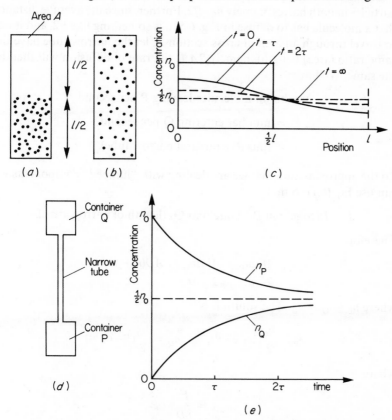

Fig. 6.5. (a) Molecules of gas initially occupying lower half of vessel, upper half being filled with another gas (not shown) to avoid convection, (b) after an infinite time, (c) concentration as a function of position at different times, (d) 'lumped' volumes and a tube, (e) concentrations in P and Q as function of time.

Without doing an exact analysis, some insight into this result can be gained from a crude model of the process. Imagine both halves of the vessel replaced by containers P and Q (Fig. 6.5(d)) one filled with the same total number of molecules as before and the other empty. Let these two be connected by a narrow tube which allows diffusion to take place. In the language of the electrical engineer we have replaced the distributed capacitance and conductance of Fig. 6.5(a) by lumped capacitances and a pure conductance in Fig. 6.5(d). (This language becomes even more appropriate if we translate the diffusion problem into the heat conduction problem, when heat capacities and thermal conductances are used.)

To make the setups comparable, let us put the same total number of particles in both halves, namely $n_0 Al/2$. Further, since the average distance that a molecule has to diffuse in Fig. 6.5(a) is something like $l/2$, and it has to travel through an area of cross section A, let the narrow tube have the same ratio (area)/(length), namely $2A/l$; the rate of diffusion will then be the same. We have:

number of molecules leaving P per second

= number entering Q per second

= (flux J) × (area of narrow tube).

To the approximation that we are dealing with "lumped" components we can use Eq. (6.1) so that

$$J = D(\text{conc'n in P} - \text{conc'n in Q})/(\text{length of narrow tube}).$$

Therefore

$$-\frac{Al}{2}\frac{dn_P}{dt} = \frac{Al}{2}\frac{dn_Q}{dt} = \frac{AD(n_P - n_Q)}{l/2}$$

where $n_P + n_Q = n_0$. The solution is

$$n_P = \frac{n_0}{2}(1 + e^{-t/\tau}), \qquad n_Q = \frac{n_0}{2}(1 - e^{-t/\tau})$$

where

$$\tau = \frac{l^2}{8D}.$$

Thus the concentrations in P and Q approach their final values exponentially, Fig. 6.5(e), with time constant $l^2/8D$ which is not very different from the $l^2/\pi^2 D$ quoted before.

This square law can have quite startling effects. Diffusion coefficients of small molecules in liquids like water at ordinary temperatures are of the order of 10^{-3} cm^2 s^{-1}. Given a tube 1 cm long, concentrations will tend to equality in times of the order of 20 minutes. But the time for a 1 m tube would be reckoned in months and for a 10 m tube it is decades. A famous example of this is a very tall vertical tube, fixed to the wall of a lecture theatre in Glasgow University. Eighty years ago it was filled by Lord Kelvin, the lower half with blue-green copper sulphate solution and the upper half with water. It is still very far from uniform in concentration.

6.2.1   Measurement of the diffusion coefficients of gases

Diffusion coefficients of one gas through another (or with certain corrections, of a gas through itself), can be measured using similar arrangements. In the experiments of Mifflin and Bennet to measure the D of argon through argon at room temperature at very high pressures for example, two volumes V, each 36 cm³, were connected by a bar of length $l = 3.8$ cm made of porous bronze (Fig. 6.6(a)). The pores were of average diameter 2×10^{-4} cm.* The total cross-sectional area of the pores was $A = 0.36$ cm², about $\frac{1}{4}$ of the area of the bar. One volume, which we will call Q, was filled with ordinary argon, the other, P, with argon at the same pressure containing a small concentration of ^{37}A. This is a radioactive isotope whose presence could be detected by the ionization it produced— each volume V was in the form of an ionization chamber in which the current was porportional to the concentration of ^{37}A. The half-life of the ^{37}A was large compared with the time taken by the experiment.

Using the same notation as in section 6.2, the equations are

$$ -V\frac{dn_P}{dt} = V\frac{dn_Q}{dt} = \frac{AD(n_P - n_Q)}{l} $$

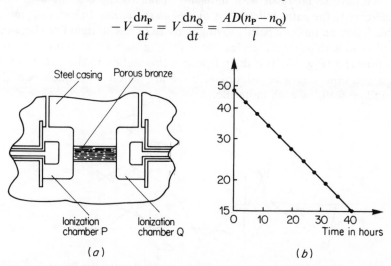

Fig. 6.6. (a) Apparatus to measure diffusion of gases at high pressures. (b) Difference of currents in P and Q in units of 10^{-14} amp, plotted on log scale, as function of time on a linear scale.

* It is shown later in this chapter that it is important that the diameter of the pores should be greater than the mean free path between collisions in order to measure D correctly. This condition was in fact satisfied here.

so that $(n_P - n_Q)$ varies as $\exp(-t/\tau)$, where $\tau = Vl/2AD$. A plot of $\log(i_P - i_Q)$ against time was, therefore, a straight line of negative slope $2AD/Vl$, the i's being the currents. Figure 6.6(b) shows the results of one run. The currents are in units of 10^{-14} amp. The slope of this graph is

$$(\ln 49 - \ln 17)/40 \text{ hours}^{-1} = (3.89 - 2.77)/(40 \times 3,600)$$
$$= 0.78 \times 10^{-5} \text{ s}^{-1}.$$

Hence $D = 1.48 \times 10^{-3} \text{ cm}^2 \text{ s}^{-1}$. This is the diffusion constant of ^{37}A in ^{40}A. A small correction must be applied in order to calculate the self-diffusion constant of ^{40}A through ^{40}A; it will be mentioned here although it can only be understood in the light of the discussion of section 6.5. We assume that the interatomic potential energies of the two kinds of atom are the same but that their mean speeds $\bar{c}$ are different, because by the equipartition law the mean kinetic energy is the same for both. This gives a 2% correction to $\bar{c}$ and hence to D. The results of these experiments are quoted in section 6.5.4.

6.3   DIFFUSION AND THE RANDOM WALK PROBLEM

We have so far dealt with diffusion in macroscopic terms with little reference to the paths followed by *individual* molecules. In fact, each molecule follows a random path, moving in more or less straight lines between collisions with other molecules, but travelling backwards almost as often as forwards (Fig. 6.7). It is the purpose of this section to show that the $\sqrt{t}$ law and other characteristics of diffusion are merely consequences of the 'random walk' of each molecule.

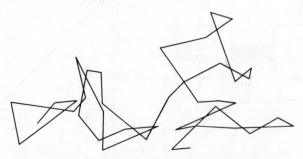

Fig. 6.7. A random path followed by a molecule, after Perrin.

The distance that a molecule travels between collisions is called the *free path*. It may be of any length, in any direction.

Let us, however, make a crude model of the random path by saying (a) that each free path is of the same length, and (b) that the molecules can only move parallel to the $+x$ or $-x$ direction—that we are dealing with a sort of 'one-dimensional gas'. It may be guessed that these simplifications allow us a considerable insight into the diffusion process although the value of the diffusivity so calculated is likely to be wrong; even then, it is not likely to be wrong by a large factor such as 10 but rather by a factor like 2 or 3.

The simplified problem is this. A molecule starts from the origin and moves a distance $\pm l$ along the x-axis; having done so, it can then move a further distance $\pm l$. Thus at the end of two such moves, it may have followed one of four possible sequences of $+l$ or $-l$ moves: namely, $(+l+l)$ or $(+l-l)$ or $(-l+l)$ or $(-l-l)$. All of these are equally likely to occur. The nett distance travelled may, therefore, be $+2l$ (which may be achieved in only one way and, therefore, has probability 1/4) or 0 (reached in two ways; probability 1/2) or $-2l$ (probability 1/4). At the end of three moves, there are eight possible sequences which may have been followed, all equally likely. The end point may be $+3l$ (1 sequence only), or $+l$ (there are 3 sequences with 2 positive-going moves and 1 negative-going), or $-l$ (again three ways of achieving this) and $-3l$ (one way only). Thus the probabilities are 1/8, 3/8, 3/8 and 1/8 respectively. Notice that the numbers 1, 2, 1 and 1, 3, 3, 1 occur as coefficients in the expansions of $(x+y)^2$ and $(x+y)^3$. Let us now generalize our results to sequences of N moves (Figure 6.8 shows two typical sequences of 10 moves, for illustration.) We can say that the total number of possible sequences is 2^N, and the number of ways of achieving Σ positive-going steps and $(N-\Sigma)$ negative-going ones is the

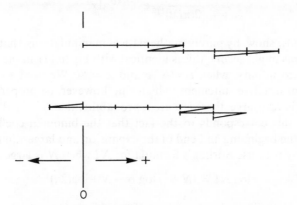

Fig. 6.8. There are 2^{10} different sequences of 10 moves, all equally probable. $10!/7!3! = 120$ of these consist of 7 positive-going and 3 negative-going moves; two of these are illustrated (slightly displaced in a vertical direction for clarity).

coefficient of $x^{\Sigma}y^{N-\Sigma}$ in the expansion of $(x+y)^N$, namely

$$\frac{N!}{\Sigma!(N-\Sigma)!}$$

These coefficients are called the binomial coefficients.

In slightly different terms, the *probability* that at the end of N moves the molecule will have travelled a nett distance $x = Sl$ (that is, S steps in the $+x$ direction), by having made $\frac{1}{2}(N+S)$ positive-going steps and $\frac{1}{2}(N-S)$ negative-going steps is

$$P = \frac{N!}{\left(\dfrac{N+S}{2}\right)!\left(\dfrac{N-S}{2}\right)!2^N} \qquad (6.11)$$

We may assume that the molecule moves with speed c during its free path so that the time required to make the N steps is $t = Nl/c$. Thus Eq. (6.11) gives the probability that a molecule will have travelled a nett distance x in time t.

We can now refer back to the problem of which Eq. (6.8) is the solution and alter it slightly to make it correspond exactly to the present one—namely by making it refer to 1 molecule instead of N_0 and to diffusion in the $\pm x$ direction instead of $+x$ only. We can interpret the solution to mean that the probablity that a molecule diffuses a distance between x and $(x+dx)$ in time t is

$$\frac{1}{2(\pi Dt)^{1/2}}e^{-x^2/4Dt}dx. \qquad (6.12)$$

We can now show, by purely mathematical manipulations, that—unlikely though this might seem—this is identical with Eq. (6.11), in the limit when N tends to infinity, when $t = Nl/c$ and $x = Sl$. We need to make one additional intuitive statement (which can, however, be properly proved), that if N is very large, there is a great probability that S is small compared with N; this corresponds to the fact that the binomial coefficients are small at the beginning and end of the expansion and largest in the middle.

The key is to use Stirling's formula for $N!$ when N is large:*

$$\log N! = (N+\tfrac{1}{2})\log N - N + \log(2\pi)^{1/2}. \qquad (6.13)$$

---

* Natural logarithms to base e are meant, of course. For many purposes

$$\log N! = N \log N - N$$

is a good enough approximation, for large N. But with all the terms present, it is remarkably accurate even for small N. It gives 10! as 3.60×10^6 instead of 3.63×10^6, for example. For $N = 10^{10}$ it is very accurate indeed.

Eq. (6.11) can be written

$$\log P = \log N! - \log\left(\frac{N+S}{2}\right)! - \log\left(\frac{N-S}{2}\right)! - \log 2^N$$

and after some tedious algebra this gives

$$\log P = \log\left(\frac{2}{\pi N}\right)^{1/2} - \left(\frac{N+S+1}{2}\right)\log\left(1+\frac{S}{N}\right) - \left(\frac{N-S+1}{2}\right)\log\left(1-\frac{S}{N}\right).$$

On the assumption that S/N is small we can write

$$\log\left(1+\frac{S}{N}\right) = \frac{S}{N} - \frac{S^2}{2N^2} + \cdots.$$

Then

$$\log P = \log\left(\frac{2}{\pi N}\right)^{1/2} - \frac{S^2}{2N}$$

or

$$P = \left(\frac{2}{\pi N}\right)^{1/2} \exp\left(-\frac{S^2}{2N}\right)$$

which in terms of distance x and times t may be written

$$P = \frac{1}{2}\left(\frac{2}{\pi clt}\right)^{1/2} e^{-x^2/2clt} 2l. \tag{6.14}$$

Comparing with Eq. (6.12), $2l$ takes the place of dx (since in this simple model there can be particles only at points separated by $2l$). The two equations are identical in form and

$$D = \tfrac{1}{2}cl.$$

Thus the essential features of the diffusion process are reproduced by this simplified 'random walk' problem. A more accurate analysis (Appendix B) gives a coefficient in the expression for D of $\tfrac{1}{3}$ instead of $\tfrac{1}{2}$; as mentioned above, this kind of discrepancy is to be expected of the over-simplified one-dimensional model.

6.4 DISTRIBUTION OF FREE PATHS

The free paths of the molecules in a gas are not, of course, all equal in length. In fact it is most probable that the free path is short and quite improbable that it should be long. The average or *mean free path* will be denoted by λ. The distribution of free paths follows the law:

$$(\text{probability of free path between } x, x+dx) = e^{-x/\lambda}\frac{dx}{\lambda}. \quad (6.15)$$

That the exponential form of this law is correct, is shown by the following argument.

(Probability of collision between $x, x+dx$)
= (probability of no collision in x) (probability of collision in dx)

and identifying the factors in this and the identical Eq. (6.15),

$$\text{probability of no collision in } x = e^{-x/\lambda} \quad (a),$$

$$\text{probability of collision in } dx = dx/\lambda \quad (b).$$

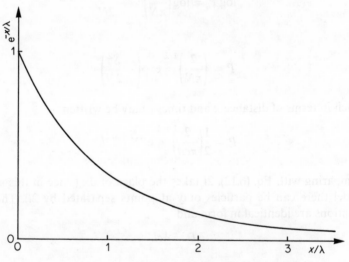

Fig. 6.9. The function $\exp(-x/\lambda)$ as a function of (x/λ).

But if (a) is correct, we would expect

$$\text{probability of no collision in } dx = e^{-dx/\lambda}$$

or

$$\text{probability of collision in } dx = 1 - e^{-dx/\lambda}$$
$$= dx/\lambda$$

for small dx. But this is exactly what we found in (b); therefore, the law is self-consistent, and satisfies the conditions imposed by the random probability of collisions. Further, it can be checked by integrating Eq. (6.15) that the probability that the free path lies between 0 and ∞ is unity, as expected; and the mean free path, given by

$$\int_0^\infty x e^{-x/\lambda} \frac{dx}{\lambda}$$

does work out at λ as required. Eq. (6.15), therefore, has all the desired properties. Figure 6.9 is a graph of $\exp(-x/\lambda)$ as a function of x/λ. It is identical in form to Fig. 3.2(b) and Fig. 4.5.

6.4.1   Mean free path and collision cross-section

Imagine all the molecules in a gas to be at rest, except one, which is moving with velocity v in a certain direction.

This molecule will collide with another if the two centres get within a distance a of one another (where a is the diameter of one molecule, although we will see in section 6.4.3 that this quantity needs careful definition). In time t, the moving molecule travels a distance vt. Any other molecules that happen to be inside a cylinder of length vt and area πa^2 will collide with it.

If there are n molecules/cm^3, the moving molecule will, therefore, make $n\pi a^2 vt$ collisions. (Of course, the molecule would be deflected but we imagine the path to be straightened out.)

Therefore, the mean distance between collisions is

$$\lambda = \frac{1}{n\pi a^2}. \tag{6.16}$$

The quantity πa^2 is called the *collision cross-section* of a molecule, denoted by σ. It is equal to 4 times the geometrical cross-section of one molecule, Fig. 6.10. We can write

$$\lambda = \frac{1}{n\sigma}. \tag{6.17a}$$

In a real gas, the molecules are not all stationary in this way, but are coming from all directions. This introduces a numerical factor: it is found that

$$\lambda = \frac{1}{\sqrt{2}} \cdot \frac{1}{n\sigma}. \tag{6.17b}$$

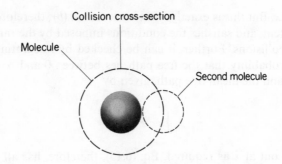

Fig. 6.10. The collision cross-section of a molecule of diameter a is πa^2. If the centre of a second molecule lies within a distance a of the centre of the first, a collision occurs.

6.4.2   Estimates of mean free paths

At 1 atmosphere pressure (that is, 760 mm of mercury), at 0°C, 1 mole of any ordinary gas occupies about 20 litres so that $n = 6 \times 10^{23}/2 \times 10^4 = 3 \times 10^{19}$ per cm³. If the diameter of a molecule is 4×10^{-8} cm, its area of cross-section is about 1.2×10^{-15} cm² so the collision cross-section $= 5 \times 10^{-15}$ cm². Thus λ is of the order of 7×10^{-6} cm, which is about 200 diameters.

The pressure of a gas is proportional to n; hence, reducing the pressure by a certain factor increases λ by the same factor. At 1/15 mm pressure, the mean free path in air is 1/15 mm—this is a useful datum for remembering the order of magnitude. At 10^{-6} mm pressure, the mean free path would be 5 m, which is bigger than most ordinary containers.

6.4.3   The dynamics of collisions

The exact calculation of the collision cross-section of the molecules of a gas is a major problem (even when purely classical laws are assumed as they will be here). If molecules were like billiard balls with a definite diameter, as we have tacitly assumed in the previous section, there would be no problem. There would be no doubt when two molecules collided and their trajectories were deflected. But with real interactomic potentials, two effects arise which have no counterparts in the billiard-ball model.

Firstly, there is an effect due to the attractive forces; this is important at low temperatures when the mean thermal energy kT is comparable with the depth ε of the potential well. The forces cause the trajectories of two molecules to bend towards one another—the paths are *correlated* in the sense that the presence of one molecule influences the trajectory of the other. Thus, two molecules which in the absence of attractions might just miss one another, may just hit. Collisions are, therefore, more frequent; the collision cross-section is increased and the mean free path decreased compared with a gas with no attractions. We will describe later, while dealing with van der Waals' equation, a crude method of allowing for this effect.

Secondly, the repulsive forces (the r^{-12} term of the Lennard–Jones potential) can be interpreted to mean that molecules act rather like elastic spheres which can be compressed together. Two colliding head-on at high speed are momentarily pushed together so that the distance apart of their centres is smaller than when the two collide at low speed. We can, there-fore, not talk in a precise way about the diameter of a molecule which behaves like this. We can, however, take the *distance of closest approach* as a measure of the effective diameter. To see how this varies with the kinetic energy of either molecule, we will state some of the more important results of Newton's laws of motion applied to collisions and then see how these apply to molecular collisions.

When two bodies collide, the centre of mass of the system is particularly important. It moves with unchanged velocity at all times during the collision. Thus, if the total mass M of the system is imagined to be concen-trated at the centre of mass, and this moves with velocity U_G, then the quantity $\frac{1}{2}MU_G^2$ is conserved—in other words, U_G and $\frac{1}{2}MU_G^2$ are un-changed, before, during and after the collision (and this holds whether the collision is elastic or inelastic). From the point of view of an observer *situated at the centre of mass and moving with it* (i.e. in a frame of reference fixed with respect to the centre of mass), an elastic collision between equal spheres looks symmetrical. The two approach one another with equal and opposite velocities along parallel paths; they are deflected, each with its speed unchanged and they go off again on parallel paths, Fig. 6.11(a).

From the point of view of *any other observer* moving with constant velocity, but not situated at the centre of mass (for example, an observer who is stationary in the laboratory, Fig. 6.11(b)), the total kinetic energy of the system is equal to the kinetic energies of the two particles with respect to the centre of mass *plus* the term $\frac{1}{2}MU_G^2$.*

During an elastic collision, some of the kinetic energy *with respect to the centre of mass* is momentarily transformed into potential energy of elastic

* This can be checked by measurement of the vectors in Fig. 6.11, which is drawn to scale. There is a well-known paradox of two trains each of mass m approaching one another, each with speed v. A stationary observer sees their kinetic energy as $2(\frac{1}{2}mv^2)$. An observer on one train sees the speed of the other as $(2v)$ and its energy as $\frac{1}{2}m(2v)^2$, which is different. The question is, how much energy is dissipated during an inelastic head-on collision. There is no paradox if it is remembered that the centre of mass of the system (stationary with respect to the earth but moving with speed v with respect to either train) conserves its speed after the collision. With respect to the observer in the train, energy $\frac{1}{2}MU_G^2 = \frac{1}{2}(2m)v^2$ is conserved and not dissipated.

deformation. For a head-on collision, all of this kinetic energy is stored as potential energy at the instant where the spheres are reversing their velocities; but for a glancing collision the tangential velocity of one sphere past the other is never zero and the spheres are less deformed.

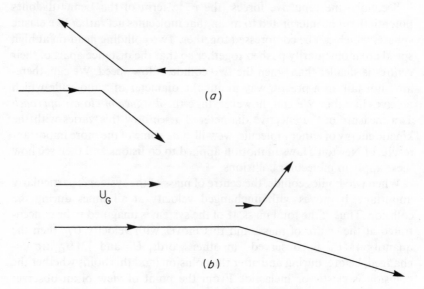

Fig. 6.11. Velocity vectors for an elastic collision of two spheres of equal mass. (a) In the centre-of-mass frame of reference, (b) in the laboratory frame of reference, where one sphere is 'chasing' the other. The velocity of the centre of mass is shown separately.

To calculate the distance of closest approach between the centres of two molecules, we therefore have to know the speeds of the two, which may have any value from zero upwards (as given by the Maxwell distribution) and their relative directions and whether the collision is head-on or glancing, Fig. 6.12(a). The *mean* distance of closest approach is governed by a complicated averaging process, and it appears that only a fraction (about $\frac{1}{4}$ or $\frac{1}{10}$) of the mean kinetic energy goes into squeezing the atoms together. It is certain, however, that the collision cross-section gets smaller with increasing temperature because the colliding atoms on average 'climb' farther up the potential energy curve. The overall result of these two effects of the interatomic potentials is that the effective collision cross-section varies with temperature as shown in Fig. 6.12(b)—which is schematic and not to be taken too literally.

It is convenient here to mention another result which is applicable to the detailed discussion of the transport coefficients. Imagine one sphere aimed at another with a given velocity, but hitting it somewhere at random so that in a large number of throws it bounces off in all possible directions. Then in the centre-of-mass frame of reference, all directions are equally probable. But in the laboratory frame, after the velocity of the centre of mass has been added to all the velocity vectors, there is a greater probability that the direction of the initial velocity will be favoured. Referring back to Fig. 4.1, for example, there is a greater probability that the molecule will get through the layer than that it will be reflected back out. This holds even if both molecules in a collision are moving and the effect is referred to as the persistence of velocities.

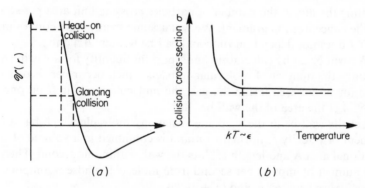

Fig. 6.12. (a) Distance of closest approach. (b) Variation of collision cross-section with temperature.

6.5   CALCULATION OF TRANSPORT COEFFICIENTS

We will now calculate the steady-state transport coefficients—diffusion coefficient, coefficient of viscosity and thermal conductivity—for gases, using the concept of the mean free path which we have developed.

It is important to notice that whenever there exists a concentration gradient or a velocity or temperature gradient in a gas, whenever any transport process is taking place, then the system is, strictly speaking, not in *equilibrium*. Conditions may be steady, but the speed distribution will not follow Maxwell's law. The exact solution of all problems of this sort then becomes difficult. Here we will assume that any departures from equilibrium and from the Maxwell speed distribution are small, that drift velocities are small compared with the velocity of sound, for example, and that any gradients are small.

6.5.1   Diffusion coefficients

We have already shown in section 6.3 that a solution of the diffusion equation

$$\frac{\partial n}{\partial t} = D \frac{\partial^2 n}{\partial x^2} \qquad (6.3)$$

can be reproduced on the simple assumption that the molecules follow a random walk.

The object of this section is to deduce the other equation

$$J = -D \frac{\partial n}{\partial z} \qquad (6.1)$$

relating the flux J (the number of particles crossing unit area per second) to the concentration gradient. We will assume steady-state conditions, so that J does not depend on time and can be written $J(z)$ only.

We will begin by calculating an important quantity for a gas in equilibrium—the number of molecular impacts which occur per second on a wall (or on an area inside the gas, the molecules coming from one side only). Let the area of the wall be A.

Let the wall be in the xy plane. Then any molecule which has a component of velocity v_z normal to it and is contained in a volume of cross-sectional area A and length v_z, hits the wall within one second. Therefore the number of impacts per second from molecules whose z-component of velocity lies between v_z and $(v_z + dv_z)$ is

$$nAv_zP[v_z]\,\mathrm{d}v_z$$

where n is the total number of molecules per unit volume and $P[v_z]$ is the probability function of v_z. (This result was written down in section 5.1.2.)

Therefore the total number of impacts from all molecules moving *towards* the wall (that is, v_z going from 0 to ∞) is

$$nA \int_0^\infty v_zP[v_z]\,\mathrm{d}v_z = nA\left(\frac{m}{2\pi kT}\right)^{1/2} \int_0^\infty v_z\,\mathrm{e}^{-mv_z^2/2kT}\,\mathrm{d}v_z = nA\left(\frac{kT}{2\pi m}\right)^{1/2},$$

using one of the integrals quoted on page 72 and the expression for the probability function from Eq. (5.3). We can rewrite this in more compact form by quoting another result, the expression for the mean speed $\bar{c}$, namely

$$\bar{c} = \frac{2}{\sqrt{\pi}}\left(\frac{2kT}{m}\right)^{1/2}. \qquad (5.10)$$

Hence it follows that the number of molecular impacts per second on an area A of a wall exposed to a gas is

$$\text{impacts/s} = \tfrac{1}{4}nA\bar{c}. \tag{6.18}$$

This result only assumes that there is equilibrium so that the Boltzmann distribution is followed.

Let us now select a small area dS normal to z in the gas—it can be imagined as a kind of little picture-frame suspended in the gas—and let us calculate the numbers of particles going through it per second, arriving from the $+z$ and $-z$ directions, Fig. 6.13. In equilibrium they would be equal, but in the presence of a small concentration gradient there is a nett flux. We will assume that near-equilibrium conditions hold.

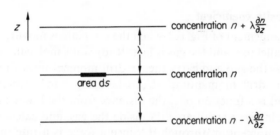

Fig. 6.13. A small area dS in a gas with molecules arriving from the $+z$ and $-z$ directions.

We will now make an approximation, which leads ultimately only to an error by a numerical factor of order unity. It is, that every molecule which passes through dS made its last collision with another molecule in the plane parallel to dS at a distance of one mean free path λ from it. This is not correct: on the average, each molecule makes its last collision a radial distance λ from dS, but not a distance λ in the z direction because many molecules travel obliquely to dS; the average distance in the z direction is something smaller than λ. Nevertheless we will make this assumption.

Let the concentration in the plane of dS be n molecules per cm^3.

The molecules coming from the $+z$ direction come from a region where the concentration is $(n + \lambda\, \partial n/\partial z)$ molecules per cm^3; those arriving from below come from a region with $(n - \lambda\, \partial n/\partial z)$ molecules per cm^3. Hence more molecules come from above than below, and the nett flux of molecules per unit area is, in the direction of increasing z,

$$J = -\frac{1}{4}\bar{c}\left(n + \lambda\frac{\partial n}{\partial z}\right) + \frac{1}{4}\bar{c}\left(n - \lambda\frac{\partial n}{\partial z}\right) = -\frac{1}{2}\bar{c}\lambda\frac{\partial n}{\partial z}.$$

which is of the same form as Eq. (6.1) expressing Fick's law. Comparing this with Eq. (6.1),

$$D = \tfrac{1}{2}\bar{c}\lambda. \tag{6.19a}$$

This is the same expression as we arrived at by considering the linear random walk*—and the error is the same. When the distribution of free paths is properly taken into account, together with the fact that the molecules can pass through dS from all angles, the result is

$$D = \tfrac{1}{3}\bar{c}\lambda. \tag{6.19b}$$

This exact expression is deduced in Appendix B.

Before comparing this result with experimental data, we will calculate the coefficients of viscosity and thermal conductivity for gases.

6.5.2 Viscosity coefficient

In the arrangement of Fig. 6.2, when the top plate is moving at constant velocity parallel to x and the gas is in a steady state, molecules continually diffuse across the gap and carry their x-drift momentum with them. Each molecule has drift momentum mU_x, where U_x is the velocity in the x direction and is a function of z, the distance from the lower plate.

Again we image a small area dS inside the gas, and calculate the nett flux of drift momentum through it. Since a force is a rate of change of momentum, the nett flux is

$$\text{Force/area} = \frac{1}{4}n\bar{c}\left[mU_0 + \lambda\frac{\partial(mU_x)}{\partial z} \right] - \frac{1}{4}n\bar{c}\left[mU_0 - \lambda\frac{\partial(mU_x)}{\partial z} \right]$$

where we have written m for the mass of one molecule, U_0 for the drift velocity in the plane of dS and we are assuming that the concentration n is constant. Hence

$$\text{Force/area} = \tfrac{1}{2}nm\bar{c}\lambda\frac{\partial U_x}{\partial z}.$$

This is identical in form with Eq. (6.7). We may justifiably alter the $\tfrac{1}{2}$ to $\tfrac{1}{3}$, and write the coefficient of viscosity

$$\eta = \tfrac{1}{3}nm\lambda\bar{c}. \tag{6.20}$$

6.5.3 Thermal conductivity coefficient

Exactly the same methods can be used to calculate the conduction of thermal energy across an area normal to a temperature gradient. The mean

* Presumably we identify the step-length l with the mean free path λ.

translational kinetic energy of molecules which collide in a region where the temperature T is given by

$$\tfrac{1}{2}m\overline{c^2} = \tfrac{3}{2}kT \tag{5.12}$$

and the nett rate of transport of energy across unit area in the direction of z increasing is

$$Q = \frac{1}{4}n\bar{c}\left(\frac{3}{2}kT_0 + \lambda\frac{\partial(\tfrac{3}{2}kT)}{\partial z}\right) - \frac{1}{4}n\bar{c}\left(\frac{3}{2}kT_0 - \lambda\frac{\partial(\tfrac{3}{2}kT)}{\partial z}\right)$$

$$= \frac{1}{2}n\bar{c}\lambda\frac{3}{2}k\frac{\partial T}{\partial z}.$$

This is identical in form with Eq. (6.4). If we again alter the $\tfrac{1}{2}$ to $\tfrac{1}{3}$ to compensate for our crude averaging,

$$\kappa = \tfrac{1}{2}nk\lambda\bar{c}.$$

This is correct for a monatomic gas for which the specific heat per *molecule* is $\tfrac{3}{2}k$. Remembering that n is the number of molecules/cm^3 we can write $\tfrac{3}{2}k$ as nC_v/N, where C_v is the specific heat per mole. ($\tfrac{3}{2}R$ J mol^{-1} deg^{-1} for a monatomic gas) or alternatively as C_v' which means the specific heat per unit volume (in J cm^{-3} deg^{-1}). For polyatomic gases, rotational energy is transported together with translational energy, so that in general

$$\kappa = \tfrac{1}{3}\frac{n}{N}C_v\lambda\bar{c} = \tfrac{1}{3}C_v'\lambda\bar{c}. \tag{6.21}$$

In words, the thermal conductivity of a gas is the specific heat per unit volume, times the mean free path, times the mean speed (which is nearly equal to the speed of sound), times a numerical factor of order unity.

6.5.4   Comparison with experiment

We have deduced expressions for the three transport coefficients for gases:

$$D = \tfrac{1}{3}\lambda\bar{c}, \tag{6.19b}$$

$$\eta = \tfrac{1}{3}nm\lambda\bar{c}, \tag{6.20}$$

$$\kappa = \tfrac{1}{3}\frac{n}{N}C_v\lambda\bar{c}, \tag{6.21}$$

where

$$\lambda = \frac{1}{\sqrt{2}}\frac{1}{n\sigma} \tag{6.17b}$$

We can compare these with the results of experiments. As may be expected, the qualitative agreement is good but the actual numbers are only correct within a factor 2 or 3.

First, a remarkable fact can be deduced if we write the viscosity as

$$\eta = \frac{1}{3\sqrt{2}} \frac{m}{\sigma} \bar{c}. \tag{6.22}$$

Here, m and σ are both constants characteristic of a given gas, and $\bar{c}$ depends only on the temperature. Thus the viscosity of a gas should be *independent of pressure*. If the number of molecules/cm³ is reduced, the mean free path increases and their product remains constant. This prediction was made by Maxwell and he undertook the experiments described in 6.1.6 to prove it. Modern measurements for argon show (Fig. 6.14(*a*)) that at 40°C it holds between 0.01 atmosphere and about 50 atmospheres pressure. The constancy is astonishing when one realizes that the left-hand side of the graph is a rough vacuum and the right-hand represent a gas at high pressure that needs a steel vessel to contain it.* The simple "billiard-ball" model seems to be adequate over this range. At extremely high pressures it breaks down, of course, Fig. 6.14(*b*). This is not unexpected—at

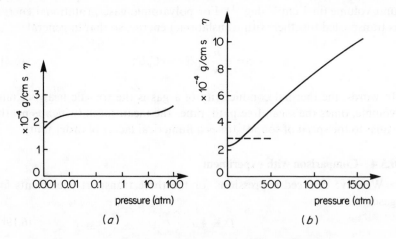

Fig. 6.14. The viscosity of argon as function of pressure (*a*) at low and moderate pressure ($T = 313°K$) and (*b*) up to high pressures ($T = 298°K$). Dotted line calculated for $\sigma = 22 \times 10^{-16}$ cm². Data from Michels, Botzen and Schuurman, *Physica* **20**, 1141 (1954).

* In a well-known demonstration experiment (the 'guinea and feather' experiment invented by Boyle) a light object, which descends only very gently when falling in air, is seen to drop much more rapidly through a rough vacuum. Yet the viscosity of the air is the same in the two cases. The paradox can be solved after reading section 9.7.1.

1,500 atmospheres and room temperature, gaseous argon is denser than water and the mean free path is less than an atomic diameter.

Similarly we can write the thermal conductivity

$$\kappa = \frac{1}{3\sqrt{2}} \frac{C_v}{N\sigma} \bar{c} \qquad (6.23)$$

where the quantities N, σ, C_v are all constants for a given gas and $\bar{c}$ depends on temperature; hence κ should be independent of the pressure. Figure 6.15(a) shows that this is so over the same wide range of pressures when η is constant—although at extremely high pressures it breaks down again, Fig. 6.15(b). This constancy at ordinary pressures is one of the most unexpected predictions of kinetic theory.

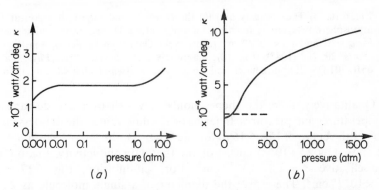

Fig. 6.15. The thermal conductivity of argon as function of pressure (a) at low and moderate pressures ($T = 313°$K) and (b) up to high pressures ($T = 398°$K). Dotted line calculated for $\sigma = 8.8 \times 10^{-16}$ cm^2. Data from Waelbroeck and Zuckerbrodt, *J. Chem. Phys.* **28**, 523 (1958), and Michels, Botzen, Friedman and Sengers, *Physica* **22**, 212 (1956).

The fact that both η and κ are proportional to $\bar{c}$ means that they should vary as $\sqrt{T}$, the square root of the absolute temperature. Figure 6.16 shows that this is roughly true. The decrease of collision cross-section at high temperature, described in section 6.4.3, is evident. It is perhaps unexpected that a gas should become *more* viscous when it is hotter; one's ordinary experience is limited to liquids, which show the opposite behaviour.

By contrast to η and κ, the self-diffusion coefficient D of a gas is directly proportional to λ by itself, and hence should be inversely proportional to the density ρ: the product $D\rho$ should be constant at all pressures. Figure 6.17 shows that this is roughly so in argon up to moderate pressures —at 300 atmospheres the density is 0.44 g/cm^3.

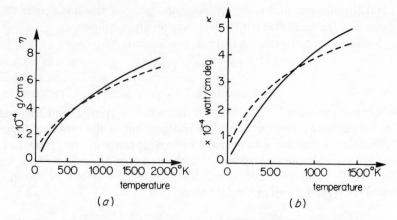

Fig. 6.16. (a) The viscosity and (b) the thermal conductivity of argon at 1 atmosphere pressure as functions of temperature. Dotted curves calculated for $\sigma = 22$ and 8.8×10^{-16} cm^2 respectively. Data from Vasilesco, *Ann Phys.* (*Paris*) **20**, 292 (1945), (Fig. 10); Kannuluik and Carman, *Proc. Phys. Soc.* **65B**, 701 (1952) and Schäfer and Reiter, *Naturwissenschaften* **43**, 296 (1956).

Qualitatively, then, the theory holds very well over a wide range of temperatures and pressures. In terms of absolute values, the fit is less good. To get the curve of Fig. 6.16(a) for the viscosity, the cross-section σ must be chosen as 22×10^{-16} cm^2. For the thermal conductivity, Fig. 6.16(b), the effective $\sigma = 8.8 \times 10^{-16}$ cm^2. For self-diffusion, Fig. 6.17, $\sigma = 16 \times 10^{-16}$ cm^2. These give the diameter of a single molecule as 2.6 Å, 1.7 Å and 2.25 Å respectively. These are certainly of the same order of magnitude as the diameter 3.35 Å, deduced from the density of the solid at absolute zero, Fig. 3.13(a); the discrepancy is in the expected direction.

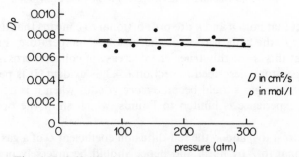

Fig. 6.17. Self-diffusion coefficient multiplied by density for argon as function of pressure ($T = 323°$K). Dotted line calculated for $\sigma = 16 \times 10^{-16}$ cm^2. Data from Mifflin and Bennett, *J. Chem. Phys.* **29**, 975 (1959).

It is perhaps not surprising that the transport processes in the gas and the density of the solid do not give exactly the same values for the molecular diameter. But the lack of self-consistence between the collision cross-sections themselves is worth trying to explain. One approach is as follows. In calculating the number of molecular impacts on an area of wall, we assumed implicitly that all directions of motion were equally likely. It is this step which is not strictly correct. It is true for a gas in equilibrium, but during transport there is some sort of drift in a special direction. In diffusion, for example, there is a nett drift of molecules in the $-z$ direction. Because of the persistence of velocity of the centre of mass of a colliding pair (see the footnote to section 6.4.3), more molecules tend to travel parallel to $-z$ after collision than in other directions; all directions are *not* equally likely before collision and so they are not equally likely after collision. Of course, this non-uniformity is small, but it is the transfer of molecules in just this $-z$ direction that we are calculating. Similar considerations apply to heat conduction. But when we deal with viscous flow, we are concerned with x-momentum being transported in the $-z$ direction (Fig. 6.2). It is the x-direction which is favoured before and after collisions inside dV, and the number passing through dS in Fig. 6.13 is altered. When the very difficult averaging processes are carried through, much of the discrepancy disappears.

6.5.5   Effusion

Suppose we have a tiny area dS which is part of a wall of a vessel containing gas. In Fig. 6.13, the xy plane can now be thought of as a wall, with the gas above it. Using Eq. (6.18), which states that when there are n molecules per cm^3 the number of impacts on unit area is $\frac{1}{4}n\bar{c}$ per second,

No. molecules hitting area dS per second $= \frac{1}{4}n\bar{c}\,dS$

$$= \frac{1}{4}\frac{P\bar{c}}{kT}\,dS$$

$$= \frac{1}{4}\frac{P}{kT}\frac{2}{\sqrt{\pi}}\left(\frac{2RT}{M}\right)^{1/2}\,dS \quad (6.24)$$

where n is the number per cm^3 and $\bar{c}$ the mean speed, and we have used the relation $P = nkT$ and Eq. (5.10) for $\bar{c}$.

Consider now a thin membrane with a hole cut in it, whose diameter is comparable with the thickness of the membrane. Let there be gas at a certain pressure on one side of this partition, and a lower pressure on the other. Gas must diffuse through the hole but we must distinguish two sets of conditions. If the diameter is large (more precisely, if it is large compared with the mean free path between collisions), any molecule suffers many collisions while going through the hole, and the description of the diffusion process as a random walk (superimposed on the bulk flow through the hole) is applicable. But if instead the dimensions are *small* compared to the mean free path—conditions which can be achieved using a tiny hole with gas at low pressure—then a typical molecule suffers its last collision some distance in front of the membrane and then goes straight through without further collision. In fact the number going through per second from one side to the other is equal to the number of impacts per second on an area equal to that of the hole. If n represents the difference of number-densities between the two sides, and P the difference of pressure, then the expressions given just about give the nett rate of transport of molecules from one side to the other.

The process of diffusion through a small hole is called 'effusion'. It is important to notice that the rate of effusion at a given temperature is proportional to $(1/M)^{1/2}$. A light gas, therefore, effuses more rapidly through a tiny hole than a heavy one.

This fact has been used as the basis of a process of great technological importance to separate gases of different molecular weights. In particular, it can be used for enriching rare isotopes found in gases consisting of mixtures of isotopes.

Imagine two compartments separated by a membrane with many small holes of suitable dimensions (Fig. 6.18). One compartment contains gas at a relatively high pressure—though it must be low enough to produce a sufficiently long mean free path—and the other side is continuously pumped to maintain a very much lower steady pressure. The 'partial pressures' exerted by the gases individually will be denoted by P_{1h}, P_{2h} on the high pressure side and P_{1l}, P_{2l} on the low pressure side.

Now the pressure of gas 1 on the low pressure side is proportional to the number of molecules effusing into it per second; similarly for gas 2. Hence

$$\frac{P_{1l}}{P_{2l}} = \frac{\dfrac{1}{2\sqrt{\pi}} \dfrac{P_{1h}}{kT}\left(\dfrac{2RT}{M_1}\right)^{1/2}}{\dfrac{1}{2\sqrt{\pi}} \dfrac{P_{2h}}{kT}\left(\dfrac{2RT}{M_2}\right)^{1/2}}$$

(where we have written P_{1h} in place of $(P_{1h} - P_{1l})$ for the pressure causing gas 1 to effuse through, and similarly for gas 2). Hence

$$\frac{P_{1l}}{P_{2l}} = \left(\frac{M_2}{M_1}\right)^{1/2} \frac{P_{1h}}{P_{2h}}. \tag{6.25}$$

This equation says that the fractional concentration of gas 1 on the low pressure side is a factor $(M_2/M_1)^{1/2}$ times as great as on the high pressure side. There is an enrichment of the lighter gas on the low pressure side.

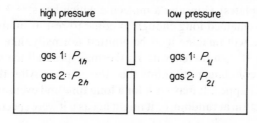

Fig. 6.18. Effusion of gas mixture through a membrane.

This is the basis of one method for enriching the rare isotope ^{235}U. Natural uranium is mostly ^{238}U with 0.7 % of the lighter isotope. From the metal, the gas uranium hexafluoride UF_6 can be produced. Its molecular weight is about 350, so the two isotopes give molecules differing by about 1 % in mass. It follows that the concentration of the lighter isotope is increased by a factor 1.005—that is, from 0.7 % to 0.7035 %—by a single passage through a membrane. The process can be made regenerative or many membranes can be used in cascade so that useful concentrations of the rare isotope can be obtained.

6.6   KNUDSEN GASES

It is not difficult to see why the viscosity and thermal conductivity of gases cease to be independent of pressure when the pressure is low (Figs. 6.14(a) and 6.15(a)). It has already been pointed out (in section 6.4.2) that at extremely low pressures the mean free path λ between collisions becomes very long. Already at 10^{-2} mm (easily attainable with a rotary pump) λ is about 1 mm for a typical gas; at 10^{-4} mm it is 10 cm. Ordinary pieces of apparatus commonly have dimensions of the order of millimetres or centimetres so that at low pressures the calculated mean free path may be larger than the apparatus. What this means in practice is that a molecule can go from one side of the apparatus to the other without making any collisions at all; the mean distance it travels is dictated by the size of the apparatus and not by the properties of the gas.

Gases at such low pressures are called 'Knudsen gases' after the scientist who first investigated them systematically. They are said to exhibit 'molecular flow' instead of viscous flow.

6.6.1   Viscous forces in a Knudsen gas

As an example of Knudsen-type behaviour, consider the force between a moving and a stationary plate immersed in a low-pressure gas (Fig. 6.19; in contrast with Fig. 6.2, there is no velocity gradient in the medium between the plates). Consider a molecule which has struck the stationary surface and remained long enough to come to rest. It jumps off—and for simplicity we will imagine it to be emitted normally, like molecule P in the diagram. It then travels with its thermal speed all the way across the gap without colliding, until it hits the other surface. After this, one of two things might happen. It may stick for a long time and eventually be emitted in some direction at random, or it might act as if it were specularly reflected.

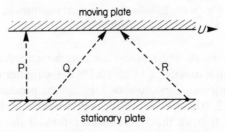

Fig. 6.19. Viscous drag with molecular flow.

The experimental evidence suggests that the molecules mostly stick and for simplicity we will assume that they all do so. Each such molecule is travelling with x-component of relative velocity $-U$ with respect to the moving plate and therefore transfers momentum $-mU$ directly to it. The same overall result holds for molecules which travel obliquely across the gap. Molecules like Q in the diagram bring some thermal momentum to the right but this is cancelled out by molecules like R, of which there are an equal number. The net result is that momentum $-mU$ is transferred from every molecule.

The total rate of transfer of momentum—the force exerted by one plate on the other—is therefore equal to the number of molecules striking the moving plate per second, multiplied by mU.

The number striking unit area per second is $\frac{1}{4}n\bar{c}$. The derivation in section 6.5.1 is independent of considerations of mean free path; the result is true as long as there is equilibrium, or near-equilibrium. Therefore the rate of transfer of momentum (the force on area A) is $\frac{1}{4}Anm\bar{c}U$.*

---

* We could arrive at practically the same result if we started from the equations

$$F = \eta \frac{AU}{d} \qquad \eta = \tfrac{1}{3}nm\bar{c}\lambda$$

and put the mean free path λ equal to the separation d.

The drag is therefore proportional to the speed of the moving plate but does not depend on the separation (provided of course that this is small). Two oscillating disc arrangements, with different spacings between the discs (Fig. 6.3) would have just the same damping; a coefficient of viscosity based on Eq. (6.7) cannot be defined. At the same time, the dependence of the viscous drag on n alone, and not on the product $(n\lambda)$, means that the force is proportional to the pressure instead of being independent of it. This behaviour is beginning to be shown at the left of Fig. 6.14(a).

For the flow of gas down a long circular tube, the analysis for viscous flow shows that the mass of gas transported per second is proportional to ϕ^4/η where η is the viscosity coefficient and ϕ the diameter of the tube. In the molecular flow region, where the effective mean free path is dictated by the diameter, the mass per second is proportional to ϕ^3.

Similar results hold for the other transport properties. Gases become thermal insulators, for example, at very low densities. An ordinary vacuum flask (dewar vessel) has double walls enclosing space evacuated to a sufficiently high vacuum for the mean free path to be limited by the spacing. Another interesting way of providing thermal insulation is to fill the interspace with a fine powder (derived from silica, and cheap to produce) which is in the form of tiny thin-walled hollow spheres, loosely packed. If the space is only roughly evacuated, the mean free path can be limited by the diameter of the spheres or the spaces between them, dimensions much smaller than the spacing between the walls of the vessel; the thermal conduction falls below the value expected if there were no powder. Powder-packed vessels are much stronger mechanically than dewar vessels; tank wagons can be insulated in this way.

APPENDIX B

B.1. Diffusion coefficient in gases

In this Appendix we calculate the diffusion coefficient in a gas taking into account the distribution of free paths—that is, not making the approximation which was made in section 6.5.1.

Imagine a small area dS, normal to the z-axis (Fig. B.1), located somewhere inside a large volume of gas. Consider a small volume dV located at distance r at an angle θ to z. We will first calculate the number of molecules which pass through dS per second, having made their last collision inside dV.

If there are n molecules/cm^3 in the neighbourhood of dV, there are n dV molecules inside this little volume.

Each one, on average, undergoes one collision every time it travels a distance λ; that is, once in every $(\lambda/\bar{c})$ seconds.

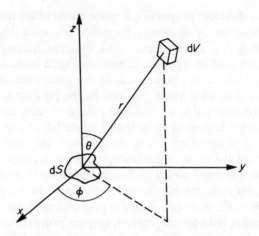

Fig. B.1. Coordinate system for calculating transport coefficients.

Therefore, the number of molecules which suffer collisions inside dV in one second is $(n\bar{c}/\lambda)\,dV$.

Each one of these molecules then goes off in some direction, and we assume that all directions are equally likely. The fraction which start off in the direction of dS is

$$\frac{\text{solid angle subtended by } dS}{4\pi} = \frac{dS\cos\theta}{4\pi r^2}.$$

Notice that by writing the solid angle in this way, we are implicitly counting molecules that come from below as negative in number. This is because when θ is greater than $90°$, $\cos\theta$ is negative. We are eventually, therefore, going to calculate the nett flux of molecules going through dS from above to below—that is, the number going downwards minus the number going upwards, the nett rate of diffusion downwards. (If we wanted to calculate the *total* number of molecules going through dS irrespective of direction we would have to take $|\cos\theta|$ instead of $\cos\theta$ in the expression for the solid angle.)

Of this fraction that start off in the direction of dS, some will suffer a collision on the way. The fraction passing straight through dS will be the fraction whose mean free path is equal to or exceeds r. Now the probability of a free path between r and $(r+dr)$ is

$$\exp\left(-\frac{r}{\lambda}\right)\frac{dr}{\lambda}$$

from Eq. (6.15), so the probability that it exceeds r, i.e., that it lies between r and ∞, is

$$\frac{1}{\lambda}\int_r^\infty e^{-r/\lambda}\,dr = e^{-r/\lambda}.$$

Collecting these results, and using spherical polar coordinates as in Fig. 5.6(b) so that

$$dV = r^2 \sin\theta\,dr\,d\theta\,d\phi,$$

we calculate that the number of molecules which collide inside dV and then travel all the way to dS and go through it in one second is

$$dS\frac{\cos\theta}{4\pi r^2}\frac{\bar{c}}{\lambda}e^{-r/\lambda}\,dV = dS\frac{n\bar{c}}{4\pi\lambda}\,d\phi\cos\theta\sin\theta\,d\theta\,e^{-r/\lambda}\,dr$$

where we are counting molecules coming from below as negative. It is convenient now to write $\cos\theta = \mu$, so that $d\mu = \sin\theta\,d\theta$. The number is

$$dS\frac{n\bar{c}}{4\pi\lambda}\,d\phi\,\mu\,d\mu\,e^{-r/\lambda}\,dr. \tag{B.1}$$

If we integrate over all space—all values of r, θ and ϕ—we take account of all molecules passing through dS, wherever their *last* collision takes place. It is obvious, however, that only values of r comparable with the mean free path λ are important, because the factor $\exp(-r/\lambda)$ means that there are a negligible number of molecules whose last collision took place far from dS.

The n which appears in this expression is the number of molecules per cm^3 inside dV. Let us now assume that there is a concentration gradient in the direction of z—that is, that n depends on z.

If $n(0)$ is the value of n at $z = 0$, the value of n at z is

$$n(z) = n(0) + z\left(\frac{\partial n}{\partial z}\right)_0 + \frac{1}{2}z^2\left(\frac{\partial^2 n}{\partial z^2}\right)_0 + \cdots$$

and higher terms are negligible because only small values of r are important. Writing $z = r\cos\theta = r\mu$,

$$n(z) = n(0) + r\mu\left(\frac{\partial n}{\partial z}\right)_0 + \frac{1}{2}r^2\mu^2\left(\frac{\partial^2 n}{\partial z^2}\right)_0 + \cdots$$

We can substitute this in (B.1) and integrate to get the nett number of molecules travelling through dS from above to below per second. To cover all possible locations of dV, r goes from 0 to ∞, ϕ from 0 to 2π and θ from 0 to π (μ from 1 to -1). If we divide through by dS, we get the nett flux

J of molecules per unit area per second from above to below. Arranging the terms:

$$J = \frac{\bar{c}}{4\pi\lambda}\left[n_0 \int_0^{2\pi} d\phi \int_0^\infty e^{-r/\lambda} dr \int_{-1}^1 \mu\, d\mu \right.$$

$$+ \left(\frac{\partial n}{\partial z}\right)_0 \int_0^{2\pi} d\phi \int_0^\infty re^{-r/\lambda} dr \int_{-1}^1 \mu^2\, d\mu$$

$$\left. + \frac{1}{2}\left(\frac{\partial^2 n}{\partial z^2}\right)_0 \int_0^{2\pi} d\phi \int_0^\infty r^2 e^{-r/\lambda} dr \int_{-1}^1 \mu^3\, d\mu + \cdots \right].$$

Let us first concentrate on the integrals with respect to μ. The first and third give even powers of μ and when limits are put in, they give zero. Only the middle term survives:

$$\int_{-1}^1 \mu^2\, d\mu = \tfrac{1}{3}[\mu^3]_{-1}^1 = \tfrac{2}{3}.$$

In this surviving term, the r integral

$$\int_0^\infty re^{-r/\lambda}\, dr = \lambda^2$$

and the ϕ integral gives 2π.

Hence the nett flux from above to below is

$$J = \frac{\bar{c}}{4\pi\lambda}\left(\frac{\partial n}{\partial z}\right)_0 (2\pi)(\lambda^2)\left(\frac{2}{3}\right) = \frac{1}{3}\lambda\bar{c}\left(\frac{\partial n}{\partial z}\right)_0.$$

This is a flux of molecules downwards, in the direction of z decreasing. To put this in exactly the same form as Eq. (6.1), we have to calculate the flux in the direction of z increasing. We must, therefore, change the sign of J. The equation is then identical with (6.1) and

$$D = \tfrac{1}{3}\lambda\bar{c}. \tag{6.19b}$$

We have performed the averaging over all directions and all lengths of free path much more precisely than in section 6.5.1, and the overall result is to justify the factor $\frac{1}{3}$ rather than $\frac{1}{2}$.

PROBLEMS

6.1. A cylindrical dewar vessel, containing water at 0°C, stands in a room where the temperature is 17°C. The glass walls are silvered to reduce heat input; the outer diameter of the inner wall is 10 cm and the inner diameter of the outer wall is 10.6 cm, the space between being filled with nitrogen gas at 1 cm pressure.
(a) Estimate the thermal conductivity of nitrogen gas.

(b) Calculate approximately the heat influx per cm height of the flask due to heat conduction.

(c) Estimate the value (in mm Hg) to which the pressure must be reduced before the heat influx begins to fall off.

(d) Deduce an approximate expression for the thermal conductivity of a Knudsen gas (compare footnote, section 6.6.1) and estimate the pressure at which the heat influx falls to $\frac{1}{10}$th of its original value.

6.2. A test-tube contains a liquid whose level falls slowly by evaporation. If the liquid is very volatile, the rate of evaporation is limited by the rate at which vapour molecules can diffuse through the air molecules. Assume that (i) the rate of fall of level is so slow that conditions are practically 'steady state' conditions (see section 6.1.1); (ii) the number of vapour molecules/cm^3 is n_v at the liquid surface and zero at the top end of the tube which is open to an infinite atmosphere.

(a) Write down the steady state equation for the number of molecules diffusing across any plane per second. Hence, the mass per second crossing any plane.

(b) Write down a simple equation giving the rate of loss of mass of the liquid in terms of the rate of fall of the liquid level, dh/dt.

(c) Hence show that the distance h of the level below the open end is proportional to $\sqrt{t}$.

(d) In a narrow tube (1 mm diameter), initially full of ether, the level falls by about 1 cm in 30 minutes, at room temperature. Deduce D for ether through air. Make a crude estimate of the diameter of an ether molecule. (Saturated vapour pressure of ether at room temperature $= 40$ cm; it obeys $PV = RT$ roughly. Molecular weight $= 74$. Density of liquid $= 0.7$ g/cm^3.)

6.3. The pressure of a gas in a thermionic vacuum tube must be such that the electron mean free path is substantially larger than the linear dimensions. If the collision cross-section of an electron with a molecule is of the order of the geometrical cross-section of the molecule, estimate the pressure required in mm of mercury. Estimate also the maximum size of pin-hole that may be tolerated in the envelope if the tube is to last for at least one year. Rate of flow (g/s) through tube of diameter D, length l, between pressure P and a much smaller pressure is $(\pi D^4/256\eta l)(M/RT)P^2$ for viscous flow and $(\sqrt{2\pi}/6)(M/RT)^{1/2}(D^3/l)P$ for Knudsen flow.

6.4. A gas at a low pressure and at a temperature T is contained in a vessel from which it effuses through a small hole whose dimensions are small compared with the mean free path. Show that the number of molecules with speeds between c and $(c+dc)$ leaving the vessel per second is $GcP[c]\,dc$ (where G is a geometrical factor which need not be evaluated. See section 5.2.4.) Hence show that the mean kinetic energy of the molecules leaving the vessel is $2kT$. This is greater than the energy, $3kT/2$, of the molecules inside the vessel. Is the equipartition law violated? Explain this result qualitatively. (It is useful to ask whether the beam is in thermal equilibrium.)

Note

$$\int_0^\infty x^5 e^{-\alpha x^2}\,dx = \frac{1}{\alpha^3}.$$

CHAPTER

7

Liquids and imperfect gases

7.1 RELATIONS BETWEEN SOLID, LIQUID AND GAS

Transitions between the solid, liquid and gaseous states of any one substance—solidification, melting, evaporation, sublimation and so on—can be brought about by varying the temperature T, pressure P and volume V. In any experiment to study these changes, the substance must (in principle) be placed inside a cylinder as in Fig. 4.2, so that the pressure acting and the volume occupied can be altered and measured. There must also be a thermostat for controlling the temperature.

The results can be displayed in several ways. In Fig. 7.1, the axes are pressure and temperature, in Fig. 7.2 pressure and volume. The lines are called phase boundary lines and represent the conditions when transitions take place. The ranges of the variables where the solid, liquid and gas phases can exist are shown as areas, on both diagrams. At the pressure and temperature represented by TP, called the triple point (a point on Fig. 7.1, a line on Fig. 7.2), all three phases can exist together.

In Fig. 7.1, a vertical line represents the course of an experiment at *constant temperature*. Three such lines are shown at low, medium and high temperature, labelled α, β and γ respectively. The same lines are shown on Fig. 7.2; they are called *isotherms*.

Let us follow the isotherm α, from low to high pressure—that is, moving upwards in Fig. 7.1 or across Fig. 7.2 starting from the low pressure, large volume region at the bottom right. At first, the substance behaves more or less like a perfect gas, obeying $PV = $ constant, so that the isotherm in Fig. 7.2 is part of a rectangular hyperbola. However, when the curve reaches the phase-boundary line, the gas suddenly begins to condense to a solid, which is of course much denser. If the piston which applies the pressure is moved so as to decrease the volume, there is no rise of pressure but more and more gas solidifies so as to keep the pressure constant. The isotherm in Fig. 7.2 therefore turns horizontal. When the substance inside the cylinder has all solidified, it becomes relatively hard to compress and the isotherm turns almost vertical.

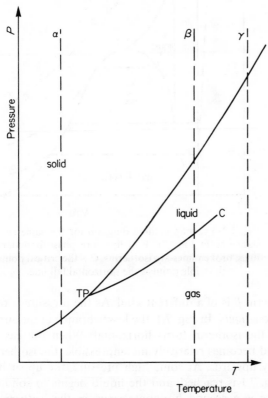

Fig. 7.1. Pressure-temperature diagram, sometimes called the phase diagram showing the relations between the solid, liquid and gas phases of the same substance. Full lines are phase-boundary lines, broken lines are isotherms. C is the critical point, TP the triple point. The boundary between solid and gas is called the sublimation curve, between liquid and gas the vapour pressure curve, and between solid and liquid the melting curve.

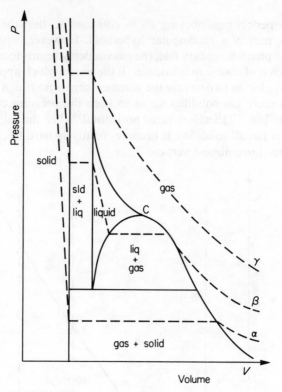

Fig. 7.2. Pressure-volume diagram for the same substance as in Fig. 7.1. Full lines are phase boundary
lines, broken lines are isotherms. C is the critical point,
the triple point is the horizontal full line.

The isotherm β is of a different kind. As the pressure is raised, the gas
condenses to a *liquid*. In Fig. 7.1, the lower branch of the curve is crossed.
In Fig. 7.2 the isotherm turns horizontal. When the gas has all been
liquefied and becomes relatively incompressible, the isotherm in Fig. 7.2
turns steeply upwards. At some high pressure, the upper branch of the
curve in Fig. 7.1 is reached, and the liquid begins to solidify, that is, to
freeze. There is a second horizontal part in the isotherm of Fig. 7.2,
corresponding to the contraction in volume from liquid to solid. When
solidification is complete, the isotherm rises steeply again. Of course this
or any other isotherm can be followed in the reverse direction; then the
solid would melt to a liquid, the liquid would boil to form the gas.

The high temperature isotherm γ is different again. The gas can be
compressed to very high density (comparable with that of the solid) before
it condenses and when it does so, it goes straight to the solid.

One isotherm, between β and γ, has a special importance. It is the one going through the point C, called the critical point. C is at the abrupt end of the lower branch in Fig. 7.1, and the top of the phase boundary curve in Fig. 7.2. This *critical temperature* is the highest temperature at which the liquid can exist.

7.1.1 Data for argon

The isotherms for argon are plotted in Fig. 7.3. Both coordinates are logarithmic, since this allows great changes of conditions to be represented

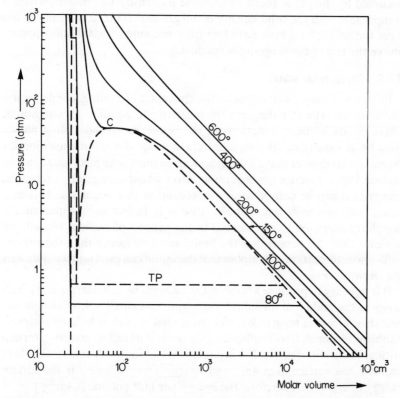

Fig. 7.3. *P, V* diagram for argon. Pressure and volume axes are on logarithmic scales. In this diagram, full lines are isotherms, broken lines are phase boundary lines. *Sources of data*: 600° isotherm: Lecocq, *J. Rech. Centre Nat. Rech. Sci.*, p. 55 (1960); 400° isotherm: Michels, Wijker and Wijker, *Physica* **15**, 627 (1949); 200° and 150° isotherms: Michels, Levelt and de Graaff, *Physica* **24**, 659 (1958); 100° isotherm: Holborn and Otto, *Z. Physik* **33**, 1 (1925). Liquid and vapour densities: Mathias *et al.*, *Leiden Comm.* **131a** (1912), and Michels, Wijker and Wijker, as above. Vapour pressures: Clark, Din *et al.*, *Physica* **17**, 876 (1951). Solid density: Dobbs and Jones, *Rept. Prog. Phys.* **20**, 516 (1957).

in one diagram. A pressure at the bottom of the graph (0.1 atmospheres or 7.6 cm mercury) is a partial vacuum; at the top, 1,000 atmospheres is an extremely high pressure. On this log–log scale, a perfect gas isotherm becomes a straight line at 45°.

The triple point is at 83.3°K, the critical temperature is 150.9°K. The 80° isotherm is of the type α, the 100° isotherm resembles β. The isotherm labelled 150° is actually just 0.2° below the critical temperature, which is why it just misses the point C. The other isotherms, at 200° and above, are of the type γ. At 200°K and 1,000 atmospheres, the molar volume occupied by the gas is about 34 cm^3 and its density 1.2 gm/cm^3, which is comparable with the solid density of 1.6 gm/cm^3. This isotherm does not meet the solid-phase boundary line till a pressure of 6,000 atmospheres, above the top of the diagram, is reached.

7.1.2 Metastable states

In this section we will consider the processes of boiling, condensation and freezing. The P, V diagram refers strictly to equilibrium conditions, where in the horizontal sections of the isotherms the coexisting phases must be at exactly equal temperatures. In a liquid–vapour transition for example, this implies that all the evaporation must take place slowly at the surface. But in practice the heat input into a liquid may greatly exceed the energy that can be carried away per second in this manner and when a liquid boils at a finite rate, bubbles arise in it. In fact, their appearance is popularly supposed to be essential to the process of boiling. We will see however that this means that the liquid must be hotter than the vapour and—more important—it implies that the liquid can exist under conditions not shown in Fig. 7.2.

It is not easy to produce a bubble. Let us begin by considering one which arises in the middle of the liquid, beginning as a small hole of atomic size and then growing bigger. We can show that if such a hole or incipient bubble is too small it will collapse again; only if its radius exceeds a certain critical value will it grow. The argument is as follows. A spherical hole of radius r has surface area $4\pi r^2$ and energy $4\pi r^2\gamma$, where γ is the surface energy in erg/cm^2. Therefore, the energy per unit volume is $4\pi r^2\gamma/\frac{4}{3}\pi r^3 = 3\gamma/r$ erg/cm^3. By arguments which have been sketched in section 3.5.1 and will be reiterated in section 7.3, we can identify this with the pressure inside the bubble—actually a more precise analysis gives $2\gamma/r$ where γ means the surface tension, in dyn/cm. This estimate is not accurate for a hole of a few atomic diameters in size where only a few bonds have to be broken but it is correct in order of magnitude even then—and the dependence on $1/r$ shows that enormous pressures must exist in order to create the hole initially. These can only be produced if the liquid is *super-*

heated above its expected boiling temperature. Let us assume that local superheating raises the vapour pressure by ΔP. Then bubbles of radius less than $2\gamma/\Delta P$ cannot be sustained and will collapse, larger bubbles can grow. Let us assume that we have argon at around 100°K, but that the conditions of heating are such that locally the temperature might reach 103°K—a moderate but not untypical degree of superheating under ordinary conditions. The vapour pressure curve gives ΔP equal to 1 atmosphere, 10^6 dyn/cm². The critical radius is then 2,000 Å, corresponding to a relatively large bubble occupying the volume of a million atoms of the liquid. Big bubbles of this kind can only be produced from smaller ones requiring greater superheating, so that it is obvious that in ordinary boiling liquids, the bubbles must be started in some other way.

In fact most bubbles form or *nucleate* at solid surfaces in contact with the liquid. The vessel which holds the liquid is likely to be hotter than elsewhere, so that bubbles would in any case be expected to form on its walls. Thin flat layers of vapour are produced and these can collect together and balloon upwards as bubbles. If the surface is rough, less superheating is needed before the bubbles can swell up in this way, and then detach themselves, though it is not at all obvious why this is so. Pieces of solid inside the liquid can also act as nucleation centres—glass seems to discharge a continuous stream of tiny particles into water so that there is never any shortage of them. Once a bubble does form, the liquid cools itself by evaporating into it and the temperature falls locally towards the equilibrium value.

Under exceedingly clean conditions however, the only way the bubbling can start is in the body of the liquid so that great superheating can occur. One can achieve this by using chemically polished tubes of newly drawn glass. More simply, it is possible to suspend liquids in oil of the same density but having a higher boiling point—water, for example, can conveniently be suspended in oil of cloves. Each liquid has a fairly well-defined temperature T_m of maximum superheating. 10° below T_m, it will last for hours; 5° below T_m it can be held for a few seconds; at T_m it explodes immediately. It will be seen later (section 7.8.1) that T_m is quite close to the critical temperature T_c.

Thus superheated liquids can be kept for long periods and studied at unexpectedly high temperatures. These are called *metastable* conditions. Instead of following the expected isotherm GADF in Fig. 7.4, comparable with the isotherm β in Fig. 7.2, isotherms like GAB can be produced. (AB does not represent the course of a superheating experiment but is a plot of the properties of the metastable liquid at constant temperature.)

Similarly, gases may be made to exist at unexpectedly low temperatures or high pressures where the P, V diagram would indicate that they should liquefy. They are called *supersaturated* vapours, and they exist because droplets have to exceed a certain critical size before they will grow; smaller droplets evaporate again. Condensation takes place most easily on rough solid surfaces or around small solid particles. In the absence of these, supercooled vapours can be kept for long periods and isotherms like DE of Fig. 7.4 can be plotted.

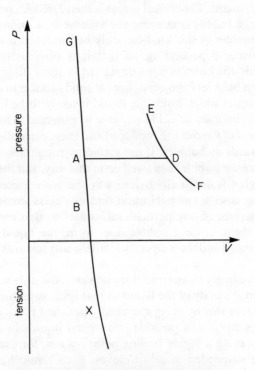

Fig. 7.4. Part of a P, V isotherm for liquid–gas
transition, showing metastable states. FDAG
is similar to part of the isotherm β of Fig. 7.2.
AB: superheated liquid. X: liquid under
tension. DE: supersaturated vapour.

Finally, liquids can be *supercooled* below their expected freezing points before they begin to solidify. Here the role of nuclei is to provide surfaces with the same crystal spacing as the solid so that this can grow by the addition of atoms to it one at a time instead of many atoms having to arrange themselves into a regular array all at once.

7.1.3   The tensile strength of liquids

The tensile strength of a substance is the tension (measured in units like dyn/cm^2) which must be applied in order to break it. For a liquid, the stretching and breaking are represented by ABX of Fig. 7.4: a liquid under tension is, of course, in a metastable state.

The tensile strength of a metal or other solid specimen can be measured by gripping its ends in some way and then pulling them apart, but this method obviously cannot be used with a liquid. Instead, the liquid must be put in some sort of tube so that when tension is applied in some way to the liquid and a bubble appears in it and it breaks, one can usually not be sure whether the break occurred inside the liquid or between the liquid and the walls. In other words, either the tensile strength of the liquid itself or the adhesion of liquid to wall may have been measured— presumably the smaller of the two.

It is observed that the tensile strength of liquids is much less than that predicted by theory. *Part* of the discrepancy may be due to the presence of tiny cracks in the walls of the tube which harbour minute bubbles of gas and prevent the liquid from entering—the very kind of nuclei which are postulated to account for ordinary boiling—and these constitute points of weak adhesion to the walls. It is indeed observed that if high *pressure* is applied for a short time to a liquid just before the tensile test begins, the observed strength is usually enhanced. This is presumably because the high pressure makes the bubbles dissolve in the liquid so that the molecules diffuse away, and do not come out again when the pressure is released.

One simple method of demonstrating the breaking of liquids is to use a hypodermic syringe to suck up, say, water through the needle. If the plunger is pulled rapidly so that the liquid rushes in, then where it leaves the narrow needle and enters the wide barrel the liquid must be under tension: bubbles can be seen to be created there. The liquid is said to cavitate. Cavitation is responsible for a good deal of noise in plumbing systems, and for great losses of energy from ships' propellors. Under controlled conditions, cavitation can be produced inside the body of a liquid, far from any solid surfaces, by passing sound waves of high amplitude through it. The breaks take place during the rarefactions. Unfortunately, this is not a good way of measuring the tensile strength of a liquid because of the large changes of temperature which also occur.

The best and most direct measurements have been taken with apparatus of the kind shown in Fig. 7.5(*a*), though without elaborate techniques for cleaning and degassing the inside of the tube the results are often a factor 10 lower than accepted values. The Z-shaped tube is made of glass and is open at both ends. It is kept in a horizontal plane by fixing it to a horizontal disc which is itself mounted at the end of a vertical shaft of a variable-speed motor. The tube is filled with liquid using a syringe, making sure that there is enough to extend round the bends into the arms. Then the tube is rotated. The liquid (surprisingly) is not thrown out but remains stably inside the tube. (This is because the centrifugal force on the liquid at A of Fig. 7.5(b) is less than at B because the distance OA is less than OB.)

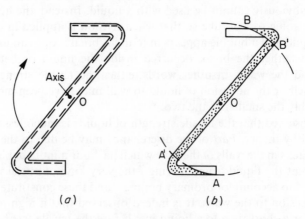

Fig. 7.5. (a) Tube for measuring the tensile strength of a liquid by centrifugal breaking. (b) Pressures at A, A', B, B' are all equal to atmospheric; hence if the liquid is originally unsymmetrical, as shown, it must move back to a symmetrical position.

Solid impurities should of course be centrifuged outwards, and bubbles, similarly, collect at the centre. At low speeds, while the nett pressure is still positive, small air bubbles sometimes collect and do not grow. It is best to stop the rotation and coax them out of the tube and to hope that some of the bigger nuclei have been got rid of.

At a distance r from the centre, a slice of thickness dr has mass $(\rho \alpha\, dr)$ where ρ is the density, α the area of cross section; the centrifugal force on it is $\omega^2 \rho \alpha r\, dr$, where ω is the angular velocity in rad/s. Each slice of the liquid exerts an outward force on the slice next to it and the total force is greatest at the centre. Integrating between the two open ends, the tension (per unit area) is $\omega^2 \rho r_0^2$, where r_0 is the radial distance to the free meniscus. The (positive) pressure of the atmosphere is added to this. The speed of rotation must be increased slowly till the liquid breaks, an event which can be easily seen because the liquid at the centre is always visible even when the tube is rotating fast. It is convenient to use a flashing stroboscope for illumination, however, because this allows both r_0 and the frequency of rotation to be measured easily.

With ordinary tap water and no special cleaning of the tube, the liquid breaks when the frequency is about 5,000–10,000 rev/min, with r_0 about 2–3 cm. This corresponds to a tensile strength of 3–10 atmospheres. With elaborate cleaning, values of the order of 10 times higher are found. The tensile strength decreases with temperature.

At room temperature, carbon tetrachloride has a tensile strength of 276 atmospheres, mercury 425 atmospheres; for liquid argon at 85°K the figure is 12 atmospheres.

7.2   THE APPROACH TO THE LIQUID STATE

It can be seen that the range of temperature over which the liquid can exist at all is a very narrow one. It is bounded at the lower end by the temperature of the triple point and at the upper end by the temperature of the point C, the critical temperature. Typically, this is only a 2:1 range in absolute temperature. By contrast, the gas can certainly exist at all temperatures, and there is no evidence to suggest that the solid cannot exist at any arbitrarily high temperature provided the pressure is high enough. This distinguishes the liquid state sharply from the solid and gas— it requires explanation.

An assembly of atoms in a rarefied gas is simple to treat mathematically, because each atom is effectively isolated and moves independently of all the others. At the other end of the scale, a perfectly regular solid can also be treated by comparatively simple mathematical methods—at any rate, if the amplitudes of vibration are not large—because all atoms are in identical, highly symmetrical environments. By contrast, the disordered array of atoms in a liquid is difficult to describe mathematically. Each atom has a large potential energy, comparable with that in a solid, due to its inter-action with a number of neighbours. But its environment is continually changing with time, and an atom is neither completely caged in by its neighbours as in a solid, nor perfectly free to move as in a gas.

Now liquids can be derived from solids by melting, and this suggests the following approach. We can consider a single atom inside an otherwise regular solid to be displaced from its lattice position, Fig. 7.6(a). This produces a small region where there is disorder, a region of high density near a hole in the lattice. When we write down the energy of such a con-figuration, the kinetic term is of course unchanged, and the potential term —the difficult one to calculate because it depends on the distances between pairs of atoms—is also not too complicated because only a few pairs of atoms are concerned. Thus we can deal with a single displaced atom. Then we can imagine many such disordered regions to be produced. Fig. 7.6(*b*). As long as they are far enough away from one another, each region will be independent of all the others and the energy required to produce *n* of them will be just *n* times the energy to produce a single one so that it can still be calculated. When the degree of disorder becomes too great, however, the problem becomes intractable. But we can in this way produce a first approximation to a liquid—or at any rate, to a solid which is showing signs of melting. Certainly, though such an assembly with a small number of displaced atoms is really quite far from being a liquid, Fig. 7.6(*c*), it can be expected to indicate relations to look for in a real liquid.

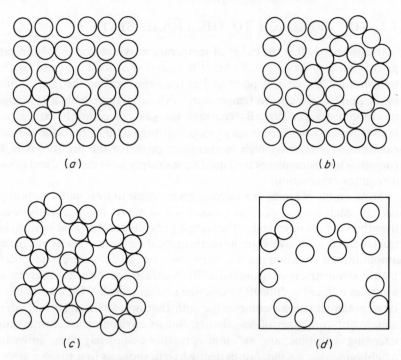

Fig. 7.6. (*a*) One atom out of place in an otherwise regular solid. (Compared with Fig. 2.5 the coordination number is low.) (*b*) Several atoms are displaced but the disordered regions are still more or less independent. (*c*) Atoms in a liquid. (*d*) A compressed gas at high density with several small clusters of atoms. Both (*b*) and (*d*) begin to resemble (*c*).

Alternatively, we can concentrate on the fact that liquids can be derived from gases by condensation. We can, therefore, begin with a gas at very low pressure and imagine the atoms brought closer and closer together so that they spend progressively more of their time near to one or more neighbours and the potential energy of the clusters of atoms cannot be neglected. The gas is then said to be imperfect. Again, only an assembly with a small number of small clusters of atoms can be dealt with simply, but even this is a rough approximation to the onset of condensation and the formation of the liquid, Fig. 7.6(d).

In Chapter 9 we will be concerned with imperfect solids, the onset of melting, and liquids in so far as they resemble solids. In this chapter, we will deal with imperfect gases, the beginning of condensation, and liquids as derived from gases.

7.2.1   Laws of corresponding states

We have emphasized the difficulties of writing down the potential energy term in liquids and have stressed that the procedure of regarding them as derived from imperfect gases cannot be pushed too far. We then meet a paradox. The simplest considerations allow us to derive an equation —van der Waals' equation of state—whose basis is particularly crude, but which is found to predict many properties of liquids themselves with surprising accuracy. By rights, van der Waals' equation should only succeed in describing the behaviour of gases which are still far from condensing; yet it is capable of predicting, in order of magnitude, such properties as the latent heat of the liquid.

There is no doubt that this surprising power is in part due to the felicitous analytical form of the equation, which is simple but adaptable. But the important fact is that once the interaction energy ε between a pair of atoms is known, many of the properties of all the phases can be estimated, as we have already seen. In van der Waals' equation, we express energies not in terms of ε itself but in terms of the critical tempera- ture T_c (or rather kT_c), which is directly related to ε. The pressure and volume at the critical point C are also taken as standard parameters.

Using the simple approach of Chapter 3, we were able to estimate the equilibrium properties of substances only at very low temperatures. Van der Waals' equation, however, suggests that we can legitimately *compare* the properties of two substances at their critical points—or at *some given fraction* of their critical temperatures, or under corresponding pressures or volumes. Conditions of this sort are called *corresponding states*.

The real significance of van der Waals' equation is therefore that it predicts certain laws of corresponding states. As a real description of phase changes, it fails badly—it does not predict the existence of the solid, for example—and even to make it refer to liquids is to use a wild and un- justifiable extrapolation. Nevertheless, more accurate theories of con- densation are extremely difficult to construct, and do not lead much further. With these caveats, van der Waals' equation will now be derived.

7.3 VAN DER WAALS' EQUATION

The object of the section is to develop a simple theory of imperfect gases and their relation to liquids. This means that we will try to explain as much as possible of Fig. 7.7, which has been dissected out of Fig. 7.2.

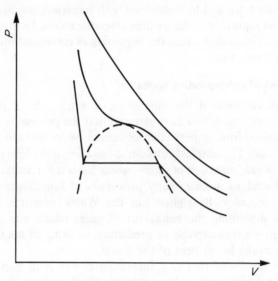

Fig. 7.7. P, V isotherms for gas and liquid, extracted from Fig. 7.2.

For an imperfect gas, we have to take some account of the potential energy of the atoms due to their interactions. These are of the form given by Fig. 7.8(a) which is identical with Fig. 3.4. r is the distance between the centres of two atoms.

Since our theory will perforce only be crude, we will simplify the algebra by substituting a rough approximation for this curve—the *square-well potential* of Fig. 7.8(b). This has the form

$$\text{Potential energy } \mathscr{V}(r) = \infty \qquad r < a$$
$$= -\varepsilon \qquad a < r < \alpha a \qquad (7.1)$$
$$= 0 \qquad \alpha a < r.$$

α is a number greater than 1 which presumably ought to be chosen so that the volume integral of the potential energy,

$$\int_a^\infty \mathscr{V}(r) . \, 4\pi r^2 \, dr$$

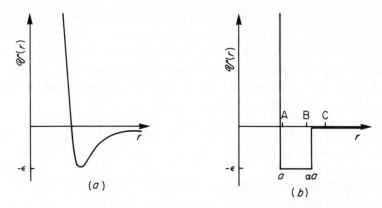

Fig. 7.8. (a) Interatomic potential energy as function of distance between the centres of two atoms; identical with Fig. 3.4. (b) Square-well potential energy, an approximation to (a), adequate for many purposes. A, B and C correspond to Fig. 7.9.

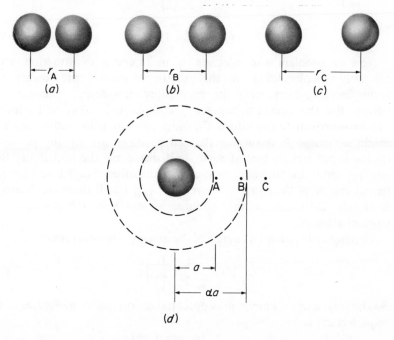

Fig. 7.9. Interactions of molecules (each of diameter a) having square-well potentials. (a) r_A is just greater than a and the system has energy $-\varepsilon$. (b) r_B is just less than αa and the energy is again $-\varepsilon$. (c) r_C is just greater than αa and the energy is zero. (d) Shows the centre of the second atom in relation to the first; the interaction volume v_i lies between the two dashed outlines.

has the same value for the 6–12 and the square well potentials.* In this way one obtains

$$\alpha^3 - 1 = \tfrac{8}{3} \quad \text{i.e.} \quad \alpha = 1.54.$$

The potential implies that the atoms are incompressible spheres of diameter a; the centre of two spheres cannot therefore approach more closely than a. At distances between a and αa, the potential energy is $-\varepsilon$, as shown in Fig. 7.9(a) and (b). At distances greater than αa, there is no interaction, Fig. 7.9(c). Another way of describing this is to imagine each atom surrounded by an 'interaction volume', bounded by spheres of radii a and αa, as in Fig. 7.9(d), so that if the centre of another atom lies inside this volume the energy is $-\varepsilon$. The size of this interaction volume is

$$v_i = \tfrac{4}{3}\pi a^3(\alpha^3 - 1). \tag{7.2}$$

If we write v_0 for the volume of a single molecule, then we have

$$v_0 = \frac{4}{3}\pi\left(\frac{a}{2}\right)^3$$

and

$$v_i = 8(\alpha^3 - 1)v_0. \tag{7.3}$$

Now the problem is to calculate the mean energy of attraction of the assembly of molecules. To do this exactly, we should write down Boltzmann factors representing the probability of different values of the energy. But this becomes impossibly complicated, so we will adopt a simple approximate procedure. Consider two atoms located somewhere inside a volume V. Assuming that all positions are equally probable (which is not strictly true at such short distances), the probability that one lies within the interaction-volume of the other is (v_i/V), and the potential energy of the pair is, on average, $-(v_i/V)\varepsilon$. If there are N atoms randomly distributed in this volume, we can select $\tfrac{1}{2}N(N-1) \sim \tfrac{1}{2}N^2$ pairs of atoms.

Therefore the potential energy of the assembly is, on average:

$$\mathscr{V} = -\frac{1}{2}N^2\left(\frac{v_i}{V}\right)\varepsilon. \tag{7.4}$$

So the total energy, kinetic plus potential energy, for N molecules of the imperfect gas is

$$\mathscr{K} + \mathscr{V} = \frac{3}{2}NkT - \frac{1}{2}N^2\left(\frac{v_i}{V}\right)\varepsilon. \tag{7.5}$$

In a perfect gas, the second term is zero.

* We identify a with the a of Eq. (3.5).

Now we have already mentioned, in section 3.5.1, that an energy density or an energy per unit volume is equivalent to a pressure, although if we use the kinetic plus potential energy this relation is only true for adiabatic changes when no heat flows into the system. But for rough calculations, the error introduced is not too serious. For example, for a *perfect* gas the energy is all kinetic and has the value $\frac{3}{2}NkT$ so that the rough rule would give $P = \frac{3}{2}NkT/V$, or in other words $PV = \frac{3}{2}RT$, instead of RT. Here, for the imperfect gas, we are calculating the effect of a small additional potential energy term, so it will serve our purpose to get the major term in the pressure correct (by omitting the factor $\frac{3}{2}$) and use the rough rule for the extra term; this is likely to introduce a numerical error but to leave the *form* of the expression correct. Then, deliberately leaving the denominator a little vague,

$$P = \frac{RT - \frac{1}{2}N^2(v_i/V)\varepsilon}{\text{volume}}. \tag{7.6}$$

In this way we have accounted roughly for the attractive forces between the atoms. If we seek a dynamical interpretation of the reduction in pressure, it is that the atoms spend more time near one another than they would if there were no attractive forces, and this reduces the number of impacts on the walls.

There is a second effect due to the interatomic potential energy, this time the repulsions which give the atoms their finite size. The volume available to the atoms to move freely about in is not V but something smaller. We should therefore, write the denominator on the right hand side as $(V - b)$, not V, where b is a volume presumably of the same order of magnitude as the volume of all the molecules.

It is not easy to decide the value of b exactly. One extreme argument is to imagine all the molecules gathered together in one lump—of solid or liquid—except for one single one, which would then obviously find a volume Nv_0 not available for moving about in, where v_0 is the volume of a single atom, $\frac{4}{3}\pi(a/2)^3$. Another extreme argument is to say that the centres of two gas molecules cannot approach more closely than a, so that a volume $8v_0$ is excluded around each molecule, making b eight times bigger than before. But this can be rejected as an overestimate because for a grazing collision between two molecules the trajectories are hardly deviated so that the centres come within just the same distance of one another as they would have done if the atoms were points; there are many more glancing than head-on collisions. Presumably, then, b lies somewhere between Nv_0 and $8Nv_0$. The most detailed analyses, taking proper account of collision dynamics show that $4Nv_0$ is the best estimate of b.

It is convenient to write $\mathfrak{a}$ in place of $\frac{1}{2}N^2 v_i \varepsilon$. It follows immediately from the connection between v_i, the interaction volume, and v_0, the volume of a molecule (Eq. 7.3) that

$$\mathfrak{a} = N(\alpha^3 - 1)\mathfrak{b}\varepsilon \tag{7.7}$$

where α is the measure of the range of the interatomic forces as shown in Fig. 7.8(b), and ε is the depth of the potential well. We have also seen that $(\alpha^3 - 1)$ is about 8/3 so that $\mathfrak{a}/\mathfrak{b} \approx 2.7\, N\varepsilon$. We have also seen, in section 3.3.1, that the molar binding energy of the condensed phase at low temperatures is $L_0 = \frac{1}{2}Nn\varepsilon$ (where n is the coordination number, about 8 or 10). Hence roughly,

$$\mathfrak{a}/\mathfrak{b} \approx \tfrac{1}{2}L_0. \tag{7.8}$$

Eq. (7.6) now reads

$$P = \frac{RT - \mathfrak{a}/V}{V - \mathfrak{b}} \tag{7.9}$$

$$\approx \frac{RT}{V - \mathfrak{b}} - \frac{\mathfrak{a}}{V^2}$$

since the second term is already a small correction;

so that $\qquad\qquad \left(P + \dfrac{\mathfrak{a}}{V^2}\right)(V - \mathfrak{b}) = RT. \tag{7.10}$

This is van der Waals' equation of state. Eq. (7.9) differs from (7.10) in one important respect: (7.9) is quadratic in V, (7.10) is cubic, and this latter form is essential to the ability of van der Waals' equation to interpret many phenomena.

We can plot P, V isotherms by assigning a value to T and calculating P as a function of V as in Fig. 7.10(a). These isotherms are of two kinds. At high temperatures, P decreases monotonically with V while at low temperatures the curves have maxima and minima, as expected of a cubic equation. We identify these two regions as above and below the critical temperature. It is obvious that the S-shaped curves are unphysical, but it is tempting to identify parts of them as representing the supersaturated vapour and the superheated liquid, and to draw horizontal lines across to represent the equilibrium mixtures, Fig. 7.10(b). No good reason for doing this can be adduced from the mode of derivation—the S-shapes are merely a result of imagining only pairs of atoms being near to one another—but if this procedure is accepted, then plausible (though not rigorous) arguments suggest that the line should be drawn so as to make the area under it equal to that under the curve. In other words, the areas of the two loops should be equal.

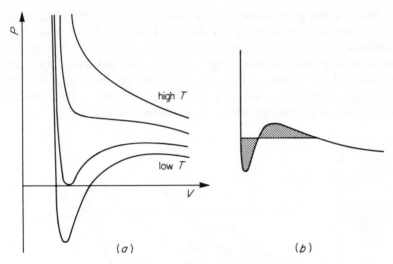

Fig. 7.10. (a) van der Waals' isotherms. (b) Horizontal line drawn so as to
equalize the areas of the loops.

7.4   APPLICATION TO GASES

We have stressed that van der Waals' equation should be valid only when
the density is not too great. We will therefore discuss its application to
gases at low and moderate densities where the equation is applicable and
at high densities where, as expected, it breaks down. We can regard a and b
in the equation as constants whose values can be chosen to fit experimental
data. It will emerge that values of a and b can be chosen which allow
several gas phenomena to be correlated. But when we compare these
values of a and b with those expected from independent estimates of the
sizes of molecules and the depth of the potential well, the agreement is in
order of magnitude only—in particular, the value of b differs by about a
factor 3 from that expected from the solid density.

Finally we will consider briefly how the theory could be improved.

7.4.1   The second virial coefficient $B(T)$

One of the most powerful methods of displaying the way a real gas
deviates from a perfect gas in its behaviour is to plot the ratio PV/RT as a
function of increasing pressure or decreasing volume. There are theoretical
reasons for preferring $1/V$ as the variable, and curves for argon are shown in
Fig. 7.11. Each curve refers to a fixed temperature. They are called virial
plots.

Most of this graph refers to small volumes—this is the effect of using $1/V$ as the variable. When V is infinite, $1/V$ is zero. When V is 100 cm^3, which is quite small compared with normal conditions, $1/V$ is only one-third the way along the axis in Fig. 7.11. The molar volume of liquid argon is about 30 cm^3 so that all the curves must be asymptotic to $1/V = 0.033$. At large volumes, that is when $1/V$ tends to zero, $PV/RT = 1$ always, so that all the curves go through one point on the vertical axis.

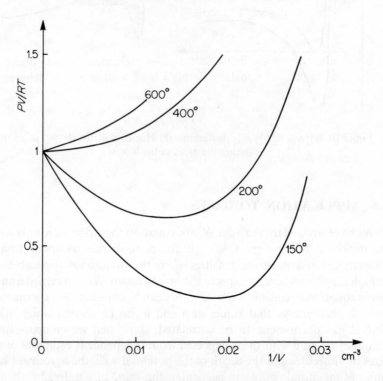

Fig. 7.11. Virial plots for argon. Sources of data as for Fig. 7.3.

Many reasonable curves $y = f(x)$ can be represented by a polynomial, $y = \alpha + \beta x + \gamma x^2 + \cdots$. Here, each virial curve can similarly be represented by

$$\frac{PV}{RT} = 1 + \frac{B}{V} + \frac{C}{V^2} + \cdots.$$

The coefficients $B, C \cdots$ are called the second, third and higher virial coefficients. They depend only on the temperature so that they should be written $B(T)$, $C(T)$ and so on—in general,

$$\frac{PV}{RT} = 1 + \frac{B(T)}{V} + \frac{C(T)}{V^2} + \cdots. \qquad (7.11)$$

For the polynomial $y = \alpha + \beta x + \gamma x^2 \cdots$, the *gradient* of the curve when x is small is equal to β (since higher terms in the expression for the gradient are negligible). Thus the gradients of the virial plots when $1/V$ is small, the initial gradients of the curves, are equal to $B(T)$. A graph of B as a function of T for argon is given in Fig. 7.12. At low temperatures, when the virial curves start downward, B is negative. At temperatures near 410°K for argon, the virial curve starts horizontally, so that $B = 0$. Around this temperature, over a considerable range of pressures, the gas obeys the perfect gas law $PV = \text{constant}$ (Boyle's law) with accuracy, whereas at other temperatures it deviates significantly at much smaller pressures. This is therefore called the Boyle temperature, denoted by T_B. At high temperatures where the curves start upwards, B is positive and tends to a constant value.

$B(T)$ can be determined from experiments while PV/RT does not deviate too much from unity—under conditions, in fact, when van der Waals' equation should be valid. We will, therefore, rearrange (7.10) as a polynomial in the form of Eq. (7.11):

$$P = \frac{RT}{V - b} - \frac{a}{V^2}$$

so

$$\frac{PV}{RT} = \left(1 - \frac{b}{V}\right)^{-1} - \frac{a}{RTV}$$

$$= 1 + \left(b - \frac{a}{RT}\right)\frac{1}{V} + \frac{b^2}{V^2} + \cdots. \qquad (7.12)$$

and comparing with Eq. (7.11)

$$B(T) = b - \frac{a}{RT}. \qquad (7.13)$$

Van der Waals' equation therefore predicts that at very high temperatures $B(T)$ tends asymptotically to b, while at low temperatures $B(T)$ becomes large and negative, following a rectangular hyperbola. The Boyle temperature T_B is evidently a/Rb.

The detailed course of $B(T)$ as calculated agrees quite well with experiment, as shown by the dashed curve of Fig. 7.12. This has been drawn with

$b = 42 \, cm^3$ and $\alpha = 1.42 \times 10^{12} \, erg \, cm^3/mol$—values which give an adequate fit over the whole range and also give α/Rb close to the observed value, 410°K, of the Boyle temperature.

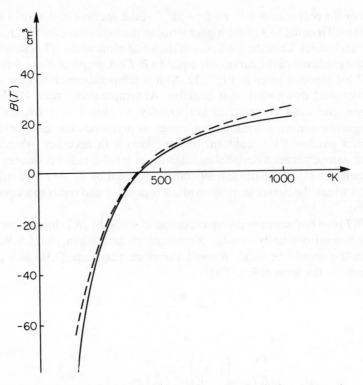

Fig. 7.12. Second virial coefficient $B(T)$ for argon. Dashed curve: van der Waals' curve calculated with $\alpha = 1.42 \times 10^{12} \, erg \, cm^3/mol$, $b = 42 \, cm^3$: $B = [42 - (1.71 \times 10^4)/T] \, cm^3$. Data from Lecocq, *J. Rech. Centre Nat. Rech. Sci.*, p. 55 (1960).

We can now apply a stringent test of the whole theory as constructed so far, by testing the prediction that

$$\alpha/b \approx \tfrac{1}{2}L_0 \qquad (7.8)$$

where L_0 is the binding energy at low temperatures. Here $\alpha/b = 3.4 \times 10^{10} \, erg/mol$, compared with $L_0 = 7 \times 10^{10} \, erg/mol$ given in Fig. 3.13(*b*). The agreement is remarkable.

The absolute value of b is not very good, however. Since we have seen that b should be about 4 times the volume of the molecules it would imply a molar volume of about 10 cm^3 for the solid instead of 26 cm^3. A plausible reason for believing that the value of B at high temperatures does never-theless depend on the volume of the molecules is provided by substances such as helium where the a term is small. Following the arguments of section 6.4.3, the value of b for a real gas would be expected to decrease at high temperatures because the molecules are not really hard spheres. In argon however, any small decrease of the b term in the expression $(b - a/RT)$ is swamped by the change in the a/RT term. But in helium where a is small, a small decrease in $B(T)$ at high temperatures is evident in the measurements—although the absolute value of b remains too small, even then.

7.4.2   Specific heats of imperfect gases

The specific heat C_v of a gas obeying van der Waals' equation is the same as if the interaction terms a and b were removed and the gas became perfect. This is because during a heating at constant volume, the mean distance between molecules (and hence the potential energy) remains unchanged. Alternatively we can argue that since

$$C_v = \left(\frac{\partial \bar{E}}{\partial T} \right)_v \tag{5.15}$$

(where we have written $\bar{E}$ for the mean value of the total kinetic plus potential energy) and

$$\bar{E} = \frac{3}{2}RT - \frac{a}{V} \tag{7.5}$$

for one mole of an imperfect monatomic gas, $C_v = \frac{3}{2}R$ as for a perfect monatomic gas.

But if a gas is heated and expands to keep the pressure constant, a quantity of heat $C_p \, dT$ must be supplied to raise the temperature by dT—not only to increase the kinetic energy of the molecules and to supply the work done, but also to increase their interatomic potential energy. For a monatomic gas:

$$C_p \, dT = d\left(\frac{3}{2}RT - \frac{a}{V} \right) + P \, dV$$

$$= \frac{3}{2}R \, dT + \frac{a}{V^2} \, dV + P \, dV. \tag{7.14a}$$

In general, for a gas whose specific heat at constant volume is C_v,

$$C_p \, dT = C_v \, dT + \left(P + \frac{a}{V^2} \right) dV. \tag{7.14b}$$

Here, dT and dV must be related because the pressure must be constant. We can find this relation as follows. Since

$$\left(P+\frac{a}{V^2}\right)(V-b) = RT, \qquad (7.10)$$

small variations of P, V and T must obey

$$\left(P+\frac{a}{V^2}\right) d(V-b)+(V-b)\, d\left(P+\frac{a}{V^2}\right) = R\, dT$$

$$\therefore \quad \left(P-\frac{a}{V^2}-\frac{2ab}{V^3}\right) dV+(V-b)\, dP = R\, dT.$$

We can neglect the term in ab/V^3. In the present case, P is held constant, $dP = 0$, so that

$$\left(P-\frac{a}{V^2}\right) dV = R\, dT.$$

This is the special relation required to substitute in (7.14b) above. After simplifying, we get

$$C_p-C_v = R\frac{P+a/V^2}{P-a/V^2}$$

$$= R\left(1+\frac{2a}{RVT}\right)^*$$

to sufficient accuracy. Thus for monatomic gas obeying van der Waals' equation

$$C_v = \tfrac{3}{2}R; \qquad C_p = \frac{5}{2}R+\frac{2a}{VT}; \qquad \gamma = \frac{C_p}{C_v} = \frac{5}{3}+\frac{4}{3}\frac{a}{RTV}. \quad (7.15)$$

This means that the ratio γ of specific heats is no longer a constant even if the classical equipartition of energy holds. γ should increase as the density and temperature are decreased, and this variation allows us to measure a again.

The ratio γ was introduced into the discussion of perfect gases (section 5.4.2) because their adiabatic elasticity is equal to γP, and the speed of sound is equal to $\sqrt{(\gamma RT/M)}$; this allows γ to be measured easily, which in turn allows both C_p and C_v to be calculated. For imperfect gases, however,

---

* The same result can be deduced with more sophistication using the relation stated in section 5.4.1:

$$C_p-C_v = [P+(\partial \bar{E}/\partial V)_T](\partial V/\partial T)_p.$$

all these results require modification. For gases obeying van der Waals' equation, it may be shown that the adiabatic elasticity is equal to $\gamma P^2 V/RT$ (which reduces to γP if $PV = RT$, as expected) and the speed of sound is

$$c_s = \frac{PV}{RT}\sqrt{\frac{\gamma RT}{M}}. \tag{7.16}$$

Thus γ can still be calculated from the measured speed of sound but with an additional factor which can be read off Fig. 7.11. For example, for argon at 200°K, at a pressure such that $V = 200 \text{ cm}^3$ and $1/V = 0.005 \text{ cm}^{-3}$, Fig. 7.11 gives PV/RT equal to 0.79. The speed of sound is equal to 256 cm/s. From these data, $\gamma = 2.52$—quite a different value from the $5/3 = 1.667$ found at low densities.

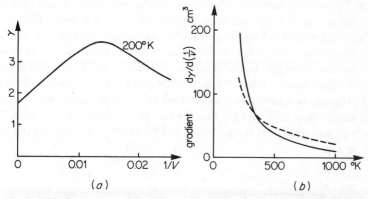

Fig. 7.13. (a) Ratio of specific heats γ for argon, as a function of $1/V$ at 200°K. When $V = 200 \text{ cm}^3$, $1/V = 0.005 \text{ cm}^{-3}$, $\gamma = 2.52$. Hence the initial gradient of this graph is $(2.52 - 1.67)/0.005 = 170 \text{ cm}^3$. Data from Michels, Levelt and Wolkers, *Physica* **24**, 769 (1958). (b) Initial gradient of this type of graph, as a function of temperature. Dashed curve: $\frac{4}{3}a/RT$; $a = 1.42 \times 10^{12}$ erg cm^3/mol. Data for gradients taken from *Nat. Bur. Std. (U.S.) Circ.*, 564 (1955).

A graph of γ as a function of $1/V$ at 200°K is given for argon in Fig. 7.13(a). γ begins by increasing linearly, as van der Waals' equation predicts, but then the curve turns downwards again when the density becomes large. The situation is analogous to that in the virial plots—van der Waals' equation predicts only the initial gradient. Proceeding as we did before, we can plot this quantity over a wide temperature range, and compare it with the predicted $4a/3RT$. The two curves are shown in Fig. 7.13(b), using the same value of a equal to 1.42×10^{12} erg cm^3 per mole as in Fig. 7.12 for the van der Waals' curve. The agreement is good. Thus even the crude representation of interatomic forces provided by van der Waals' equation is enough to account qualitatively for the specific heat variations.

7.4.3 Free expansion of gases

In section 5.4.2 we discussed the temperature change when a perfect gas (with $a = 0$ and $b = 0$) performs work by pushing a piston back—an adiabatic expansion with the performance of external work. If however a perfect gas undergoes an expansion without performing any external work, no energy is expended and there is no temperature change. This is called a free expansion and we will describe how it can be performed in principle although measurements of this kind are rarely performed in practice.

Imagine a vessel constructed of rigid material, thermally insulating and of negligible heat capacity (Fig. 7.14). Inside, it is divided into two compartments one of which contains a gas under pressure, while the other is empty. The wall dividing the compartments is then broken and gas flows so as to equalize the pressure throughout. Then, while the gas is flowing, one compartment gets hot because the gas there is being compressed more or less adiabatically, while the other side gets cold because it is doing the compressing. Imagine then that the two halves later reach equilibrium by exchanging heat with one another, and come to equal temperatures. If the gas is *perfect*, this final temperature will be exactly the same as the initial temperature.

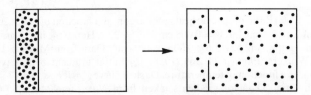

Fig. 7.14. Free expansion of a gas inside a rigid, insulated vessel.

During the whole process, the gas is isolated from its surroundings by the rigid vessel so that energy neither enters nor leaves. The total internal energy, kinetic plus potential, of the gas must be constant. Let us therefore consider one mole of an *imperfect* monatomic gas, whose energy is given by the van der Waals' expression

$$\bar{E} = \frac{3}{2}RT - \frac{a}{V} \tag{7.5}$$

where V is the volume it occupies. This is conserved. If the subscript i denotes the initial and f the final conditions, then

$$\frac{3}{2}RT_i - \frac{a}{V_i} = \frac{3}{2}RT_f - \frac{a}{V_f} *.$$ (7.17)

For 1 mole of argon expanding from 1 litre to 2 litres, that is from about 20 to 10 atmospheres, the change of temperature would be 4° (taking a to be 10^{12} erg cm^3/mol). This is a large change but in practice the heat capacity of the vessel—necessarily thick walled—would decrease its magnitude and we will attempt no comparisons with the meagre experimental data.

7.4.4   Joule–Thomson coefficient

The Joule–Thomson porous plug experiment is a much more sensitive method of measuring the change of internal energy of a gas with pressure. It is a continuous process (as opposed to a "one-shot" process like the one just described) so that the temperatures ultimately reached by different parts of the apparatus do not depend on their heat capacities.

In principle, Fig. 7.15, a gas is maintained at pressure P_1 (by an external compressor) and is brought to a known temperature T_1. It is forced through a device which can maintain a pressure difference, and does not allow heat to be conducted across it. In the original experiments, carried out in 1852, this was a silk handkerchief. Nowadays, a cotton-wool plug or a porous ceramic plug is used; often, just a long length of narrow-bore tubing with a small hole in the end. The gas then emerges into a space

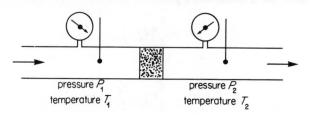

pressure P_1 pressure P_2
temperature T_1 temperature T_2

Fig. 7.15. The flow of a gas from high to low pressure through a porous plug.

* The general expression for the temperature change in a free expansion is

$$C_v(\partial T/\partial V)_{\bar{E}} = P - T(\partial P/\partial T)_v$$

which for a gas obeying van der Waals' equation is $-a/V^2$. This gives

$$C_v\,dT = -(a/V^2)\,dV$$

in agreement with the above expression.

which is maintained at another known pressure P_2. The temperature T_2 at the exit side is measured. The gas is continuously forced through the plug, and measurements are made only when the parameters are steady.

In such a process, the gas is certainly not isolated from its surroundings, so that the internal energies of 1 mole of the gas on one side and the other are not equal. However, we can still find a quantity which is conserved. First, we note that in the steady state there is no nett interchange of heat with the walls. Secondly, we can assume that the velocity of bulk movement of the gas is so small that its bulk kinetic energy can be neglected. Finally, we concentrate on the balance between the internal energy of the gas and the work performed on it or by it.

Let 1 mole of the gas occupy volume V_1 on the entrance side. The work *done on it* to force it through the plug is $\int P \cdot dV$ where P is constant at P_1 and V is changed from V_1 to zero; that is, $(P_1 V_1)$. On the other side, the same gas *performs work* $P_2 V_2$ on the pump. The nett amount of work must come from the internal energy, so that

$$\bar{E}_1 + P_1 V_1 = \bar{E}_2 + P_2 V_2.$$

In thermodynamics, $\bar{E} + PV$ is called the enthalpy, and it is this quantity which is conserved here.

Expanding van der Waals' equation and neglecting the very small term in ab/V^2:

$$PV = RT - \frac{a}{V} + bP.$$

We also have, for a monatomic gas,

$$\bar{E} = \frac{3}{2}RT - \frac{a}{V}. \tag{7.5}$$

So

$$\bar{E} + PV = \frac{5}{2}RT - \frac{2a}{V} + bP.$$

We can now calculate the temperature change dT accompanying a *small* change of pressure dP. First we can, with little loss of accuracy, substitute $PV = RT$ in the small a/V term:

$$\bar{E} + PV = \frac{5}{2}RT - \frac{2aP}{RT} + bP.$$

Since this is conserved,

$$d(\bar{E} + PV) = 0.$$

This gives

$$\frac{5}{2} R \, dT + \frac{2aP}{RT^2} \, dT - \frac{2a}{RT} \, dP + b \, dP = 0$$

$$\frac{dT}{dP} \text{ for a J–T process} = \frac{2a/RT - b}{\frac{5}{2}R + 2aP/RT^2}. \tag{7.19a}$$

Writing $2a/VT$ in place of $2aP/RT^2$, the denominator is seen to be C_p for the monatomic gas, Eq. (7.15). Generalizing, one obtains for the Joule–Thomson effect in any gas obeying van der Waals' equation:

$$\frac{dT}{dP} \text{ for a J–T process} = \frac{2a/RT - b}{C_p}. \tag{7.19b}$$

This is called the differential Joule–Thomson coefficient. Both dT and dP represent increases. In a real experiment, dP is always negative. Therefore, if $(2a/RT - b)$ is positive, the drop of pressure will cool the gas. Evidently this should occur if the temperature is low, because then the first term is large. Conversely, a heating should occur at high temperatures. The changeover from heating to cooling, at a temperature called the inversion temperature T_i, occurs when $T = 2a/Rb$. Van der Waals' equation predicts that the inversion temperature should be twice the Boyle temperature T_B.

Agreement with these predictions is surprisingly good. Figure 7.16 shows the observed differential Joule–Thomson coefficients for argon (extrapolated to zero pressures where C_p is accurately 5R/2). The observed inversion temperature $T_i = 785°K$; thus the ratio $T_i/T_B = 785/410 = 1.92$, close to the predicted value 2. The dashed curve is that predicted from van der Waals' equation with $a = 1.42 \times 10^{12}$ erg cm^3/mol, $b = 42$ cm^3, the same values used for the virial coefficients and γ curves, Figs. 7.12 and 7.13.

7.5   REFINEMENTS TO VAN DER WAALS' EQUATION

The problem of deducing an accurate equation of state, valid at all densities, is basically one of writing down the potential energy of an enormous number of interacting molecules. Van der Waals' equation takes only the first step in this direction. In the following sections we will consider some important effects which have not been taken into account, and some ways in which van der Waals' equation might be refined.

---

* The general expression is $(\partial T/\partial P)_H = \{T(\partial V/\partial T)_p - V\}/C_p$ where H stands for enthalpy. For a gas obeying a virial equation, this reduces to $\{T B'(T) - B(T)\}/C_p$ to a first approximation.

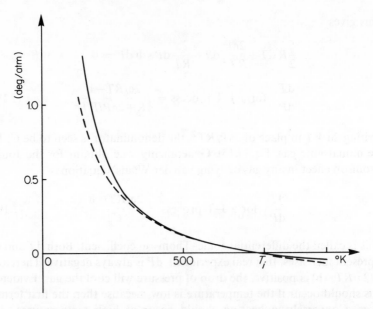

Fig. 7.16. The differential Joule–Thomson coefficient for argon as a function of temperature. Since 1 atmosphere = 1.01×10^6 dyn/cm^2, 1°K/atm ~ 10^{-6} °K/dyn.cm^{-2}. Dashed curve: van der Waals' curve calculated with $a = 1.42 \times 10^{12}$ erg cm^3/mol, b = 42 cm^3: Joule–Thomson coefficient $= \left(\dfrac{163}{T} - .203\right)$°K/atm. Continuous curve: experimental results. Sources of data: as for Fig. 7.13(a), (b).

★ **7.5.1   Dimers**

When two atoms in a gas are a small distance apart comparable with their diameters, they have an appreciable (negative) potential energy. In an ordinary encounter between two atoms, they approach one another and then fly past; it is only while they are close together that they have potential energy comparable with $-\varepsilon$. The pairs of atoms which we have considered so far are not bound permanently together in any way; they are merely pairs of atoms which happen for a short time (a fraction v_i/V of the total) to be in one anothers' vicinity.

But the existence of the minimum in the $\mathscr{V}(r)$ curve means that it is possible to form pairs of atoms which are *loosely bound together*. These are called dimers. It can be shown theoretically that the commonest form of dimer is *not* two atoms statically stuck to one another, or oscillating as if joined by a spring; instead, they *rotate* round one another—in orbit round one another, like a double star. The two molecules 'touch' one another and the mean frequency of rotation is about 10^{11} rev/s.

To form such a pair out of two isolated atoms requires rather special circumstances. As has been mentioned, two atoms which come together will usually fly apart again; but if by chance there is a third atom in the vicinity at the right moment which can take away enough kinetic energy, then the pair can be left in a bound orbit. This is called a three-body encounter. Equally well, another collision with a sufficiently energetic atom can knock them apart again. The energy required to disrupt a dimer is practically equal to ε if they are orbiting only very slowly, but it is reduced if they are orbiting fast. The 'centrifugal potential' $L^2/2I$ has to be added to the interatomic potential energy, where L is the angular momentum and I the moment of inertia of the system. There is therefore a limiting angular momentum above which the potential well is filled up and the dimer cannot exist at all.

During the whole of its existence, a dimer makes an appreciable contribution to the potential energy of the gas. At high temperatures, dimers are likely to be knocked apart again after a short time, but at low temperatures when they are longer lived, their energy can dominate the second virial coefficient $B(T)$.

Dimers have been detected and their masses measured experimentally. This was first done by Leckenby and Robbins in 1965. The basic idea was to produce a narrow beam of gas atoms, bombard it with electrons so as to produce ions and then to analyse the masses present in the beam by passing it through a mass spectrometer. The biggest technical difficulty was to produce the narrow beam inside the high vacuum required for the mass spectrometer to function. This was done by allowing the gas to escape through a tiny hole from a reservoir where the pressure was of the order of 1 mm of mercury; the thickness of the diaphragm and the diameter of the hole in it were both of the order of 10^{-4} cm, comparable with the mean free path of the molecules in the gas. Under these conditions, the beam effused through without change of temperature or mean energy, a true sample of the molecules inside the reservoir.

With argon (atomic weight 40) molecules of mass 80 were detected. They were shown unambiguously to have originated inside the reservoir and not spuriously as the result of any process inside the mass spectrometer. In the gas at room temperature at 10 cm pressure the dimer concentration was found to be 1 in 10^4, in agreement with theoretical estimates. Of course these experiments can only be conducted with low pressures in the reservoir; dimer concentrations are greatly increased at high pressures.

One unsuspected fact was revealed in these experiments. It has been mentioned in section 5.4.4 that specific heat measurements indicate that monatomic molecules in gases at ordinary temperatures do not rotate about their centres, whereas polyatomic molecules (such as N_2 or CO_2) do rotate. Now a dimer of N_2 or CO_2 has three sorts of rotation going on inside it—the spinning of each molecule about its centre and the orbiting of the molecules round one another. Within this system, angular momentum must be conserved. If therefore one of the molecules stops spinning for any reason, the orbiting must speed up—and this will probably cause the pair to fly apart. In monatomic gases however, this effect cannot occur. The result is that polyatomic gases contain fewer dimers than monatomic gases.

★ **7.5.2   Higher clusters**

The pairs of molecules which are not bound together but nevertheless possess some potential energy because they happen to be in one anothers' vicinity for a short time are called *clusters* of two. Bigger clusters are of course possible, clusters of three or more. It was thought for a long time that three molecules meeting at a point would be so rare an event that its probability could be neglected: but this is not true at high densities. Taking into account the potential energy of the higher clusters will obviously alter the equation of state. How it will do so can be guessed from the following line of argument. Let us first consider a gas of hard-sphere molecules, but with no attractive forces: that is, ε is zero but a is not zero, or in other words a is zero but b is not zero. Then the equation of state becomes

$$P(V-b) = RT.$$

Expanding this as a virial equation

$$\frac{PV}{RT} = \frac{V}{V-b} = 1 + \frac{b}{V} + \frac{b^2}{V^2} + \frac{b^3}{V^3} + \cdots.$$

In other words, for the hard-sphere model, the second virial coefficient is b, the third b^2 and so on. However, we have seen that when we put in the attractive forces and consider clusters of *two*, the *second* virial coefficient becomes $(b - a/RT)$ instead of b, while the others are left unchanged. We may therefore guess that if we consider clusters of 3 the *third* virial coefficient will be modified and so on. The higher clusters are therefore important because they allow the higher virial coefficients to be calculated properly.

We will now outline how the clusters of 3 could be dealt with, using the same rough methods as we did for the clusters of 2. The argument will not be followed through to the end because the approximations are too crude; the purpose of this calculation is to indicate some interesting features of the argument which have their analogies in more sophisticated treatments.

Consider a volume V containing a single molecule and a cluster of 2 molecules. The interaction volume of the cluster is shown in Fig. 7.17. If the third molecule enters either of the volumes v_A, it is near only one neighbour and loses energy ε only. But if it comes within the volume v_B, it interacts with *both* of them and loses energy 2ε. Thus there are two distinct types of clusters of 3.

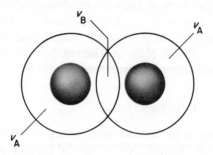

Fig. 7.17. Interaction volume of a cluster of 2.

Assuming that the distribution is still random, the probability that the third molecule finds itself within one or other of the volumes v_A is $2v_A/V$; within v_B, the probability is v_B/V. Thus the average extra energy lost by the cluster of 2 because of the presence of the third molecule is

$$\varepsilon \frac{2v_A}{V} + 2\varepsilon \frac{v_B}{V}$$

which can be written $2v_i/V$ because

$$2v_A + v_B = 2v_i.$$

We must now put in the probability that the clusters of 2 was formed originally, and the fact that in an assembly of N molecules there are $N(N-1)(N-2)/3! \approx N^3/3!$ ways of selecting three of them. The average potential energy of the assembly is thus

$$-\frac{N^2}{2!}\varepsilon\left(\frac{v_i}{V}\right) - \frac{N^3}{3!}2\varepsilon\left(\frac{v_i}{V}\right)^2.$$

We can then write down the energy density and get a small pressure term in V^{-3} as well as the V^{-2} term of van der Waals' equation. Expanded as a virial equation, the third virial coefficient can be picked out. It depends on temperature but is still of order b^2.

The problem of computing the energy of all the possible types of cluster becomes rapidly more complex with the size of cluster, if more realistic potentials are used in place of the square well. Calculations of this type have nevertheless been pursued because it was hoped that the S-shaped curves of Fig. 7.10(a) would be eliminated if all clusters could be included. In fact, dimers and higher *bound* groups of molecules probably play a dominant role in the process of condensation. They have been detected by the same kind of experiment as that described in section 7.5.1, but allowing the gas to enter the vacuum through a comparatively wide nozzle so that it cooled to a low temperature by expansion. Dimers, trimers and all degrees of association were found up to 40-molecule aggregates, the upper limit of the instrument. Some of the biggest of these might almost be thought of as small droplets.

7.6 CRITICAL CONSTANTS

Having emphasized the inadequacies of van der Waals' equation and the fact that it cannot be expected to be valid beyond moderate densities, we will nevertheless apply it to high densities. Whereas practically all the former results could be deduced equally well from Eq. (7.9) instead of Eq. (7.10), the cubic form of van der Waals' equation is now essential.

We have already noted that the isotherms of a van der Waals gas have an S-shaped form below a certain temperature, which we identify with the critical temperature.

We can select the critical isotherm by first finding the locus of maxima and minima, and then finding the maximum of *this* curve.

Starting from van der Waals' equation

$$P = \frac{RT}{V-b} - \frac{a}{V^2},\tag{7.10}$$

we find the equation of the curve on which all turning points lie by differentiating with respect to volume, keeping the temperature constant,

$$\left(\frac{\partial P}{\partial V}\right)_T = -\frac{RT}{(V-b)^2} + \frac{2a}{V^3}$$

$$= 0 \quad \text{at the turning points.}$$

We can get a more useful expression by using van der Waals' equation to eliminate RT. We get

$$\frac{2a}{V^3} = \frac{RT}{(V-b)^2} = \frac{P+a/V^2}{(V-b)}$$

or

$$P = \mathfrak{a}\frac{(V-2b)}{V^3} \qquad (7.20)$$

as the equation of the locus of maxima and minima, Fig. 7.18. The maximum of *this* curve is given by equating the gradient to zero:

$$\frac{dP}{dV} = \frac{2\mathfrak{a}}{V^4}(3b - V) = 0.$$

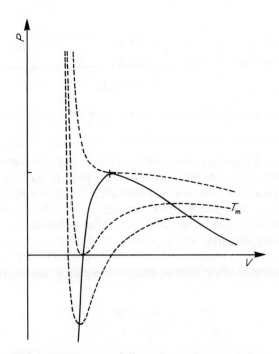

Fig. 7.18. Full curve: locus of maxima and minima of van der Waals' isotherms, Eq. (7.20). Isotherms also appear in Fig. 7.10(*a*). The isotherm labelled T_m is referred to in section 7.8.1.

Thus at the critical point:

$$V_c = 3b. \qquad (7.21a)$$

Substituting this value back in the equation for the locus of maxima and minima,

$$P_c = \frac{\mathfrak{a}}{27b^2} \qquad (7.21b)$$

and putting both these in van der Waals' equation,

$$T_c = \frac{8a}{27Rb}. \tag{7.21c}$$

It has already been shown, when a and b were originally defined, that we expect (from Eq. (7.8)):

$$a/b = \frac{8}{3}N\varepsilon.$$

This gives

$$T_c = \frac{64}{81}\frac{\varepsilon}{k}$$

$$= 0.79\,\varepsilon/k.$$

Roughly speaking therefore, the critical temperature T_c occurs when

$$kT_c \sim \varepsilon \tag{7.22}$$

that is, when kT becomes comparable to the energy of interaction of two molecules. When the thermal energy exceeds this, the gas cannot liquefy.

We could proceed to compare the critical constants of a number of gases with the values of a and b which we derive from the Joule–Thomson coefficient or from knowledge of the radii of the molecules and the form of the square-well potential. But it is more realistic to use van der Waals' equation to provide laws of corresponding states—for example, to *compare* liquids at their critical points. We can write for the van der Waals gas

$$\frac{RT_c}{P_c V_c} = R\frac{8a}{27Rb}\frac{27b^2}{a}\frac{1}{3b} = \frac{8}{3}. \tag{7.23}$$

This relation is obeyed remarkably well. For argon, $T_c = 150°\text{K}, P_c = 48.3$ atmospheres $= 49 \times 10^7$ dyn/cm^2, $V_c = 74.6$ cm^3, so that this ratio is 3.41. This is within 30% of the predicted value, 2.67. Data for other gases are:

	T_c °K	P_c atm	V_c cm^3	$\frac{RT_c}{P_c V_c}$
Nitrogen	126	33.5	90	3.43
Carbon dioxide	304	73	94	3.63
Water	647	218	56	4.35

7.7   FLUCTUATION PHENOMENA

It is convenient at this stage to introduce an important aspect of statistical theory, namely the fluctuations that occur in any statistical quantity. It is a concept which is applicable to any branch of statistics and it could have been introduced in section 4.2 where statistical ideas were discussed. However, the optical phenomena which occur at the critical points of liquids are among the most striking manifestations of density fluctuations in fluids. We will therefore concentrate on phenomena near the critical point and then extend the discussion to fluctuations in general.

7.7.1   Critical point phenomena: critical opalescence

The region of the critical point has been much studied and some peculiar effects seem to occur there. It seems that the temperatures at which the isotherm becomes horizontal, at which the meniscus disappears and at which the properties of the two phases become identical may all be different. Thus, over a narrow interval of temperature the fluid can exist in a tube as two layers of different density even though there is no sharply defined meniscus separating them. At the same time, it is known that the properties of the liquid and vapour in the region of the critical point are strongly influenced by minute traces of impurity and it is difficult to decide whether the observations are highly significant for a full understanding of the process of condensation or whether they are in some sense spurious.

One phenomenon however certainly uncovers some interesting physical ideas. The appearance of a normally colourless fluid at its critical point is remarkable. Illuminated by a beam of light, it looks diffuse and shimmering and intensely white, so that one instinctively thinks that a cloud fills the whole space. If the temperature is raised or lowered by as little as a fraction of a degree away from the critical point, the whiteness disappears and the gas or liquid appears colourless again as one thinks it should. This phenomenon is called 'critical opalescence'.

The process undergone by the light in its passage through the fluid, going in as a beam but coming out diffusely in all directions, is called *scattering*. Now it can be shown that a large block of a perfectly regular crystalline solid at absolute zero scatters no light at all—it would be invisible except for reflections from the surface—provided that the wavelength of the light is much greater than the interatomic spacing, a condition which is always satisfied. The fact that the fluid scatters so strongly near the critical point indicates that it is in some way far from homogeneous. We will be able to show (in section 7.7.3) that this is indeed so, and that its density varies appreciably from point to point at the critical point. We cannot however, give a satisfactory account of the *optical* effects, for the following reasons. It can be shown that when the density of a medium varies from point to point *in an entirely random way*, the variation being appreciable over distances small compared with the wavelength, then short wavelengths are scattered more than long ones and as a result the medium looks *blue*. The blue of the sky, for example, originates in the light scattered by the great thickness of air above one's head; this in itself is sufficient proof that the air is composed of a completely random arrangement of small molecules. Evidently, a medium where the variations of density are random but particularly large would scatter *blue* light with great intensity. But in a highly compressed gas or a liquid the molecules are not arranged completely at random. There are regions where several molecules are nearly close-packed, whose arrangement over a short distance is fairly regular. It is this fact that causes the scattered light near the critical point to be white rather than blue. Measurements of the intensity at any angle can indeed give information about the scale of distance over which the molecules are ordered—which bears a distant relation to the average size of the clusters. Experiments show that at the critical pressure but 10^{-3} deg above the critical temperature, this 'correlation length' is about 1000 atomic diameters; $\frac{1}{10}$ deg away it is about 10 diameters. We cannot follow these arguments, however, but will content ourselves by showing that the variations in density which occur near the critical point are particularly large, without attempting to estimate the intensity of the scattered light or to predict its colour.

7.7.2   Concepts of probability theory—II. fluctuations

The variations of density from point to point which occur in every system but which are very great in a liquid at its critical point are a *fluctuation phenomenon*. We will now consider fluctuations in general. It must be emphasized that they occur in all systems and that they are *not* due to gross effects like unequal heating or nonuniform external pressures: they occur in systems in thermal equilibrium. Their origin lies in certain aspects of statistical theory which we have so far ignored.

In section 4.2 we called attention to the fact that when we deal statistically with small numbers of people, the characteristics of a single individual can quite upset the shape of a histogram. Thus, Fig. 4.3(*a*) which refers to a small sample of 100 people with at most 15 in any range, is not a regularly stepped histogram; Fig. 4.3(*b*) which differs from it merely in the much larger numbers in any range, is regular. Such deviations from the most probable value occur in all types of statistical phenomena. We will, however, concentrate on physical examples.

The most probable situation inside a gas or liquid in thermal equilibrium is that (in the absence of external fields) its density is uniform at all points, on the average. But instantaneously this is not so, as can be seen by referring to the pictures of the molecular arrangements in gases, liquids and solids, Figs. 2.3, 2.4, 2.5. Liquids under ordinary conditions contain 'holes' whose size is comparable with 2 Å cubed, though there are many regions where the packing is close; and the pattern changes all the time. Obviously, then, the density can fluctuate over a wide range if we concentrate on small volumes. This can be seen by making a mask of size corresponding to 3 Å square, and laying it down over Fig. 2.4: the number of molecules encompassed varies from about 1 (near the hole in the lower right-hand corner) to about 2 wherever the packing is close. This is a 100 % variation. If on the other hand we deal with large volumes, the fluctuations get relatively smaller; a mask corresponding to 10 Å square encloses between about 20 and 24 molecules in Fig. 2.4, only a 20 % variation. It is indeed a general principle that the smaller the average number of molecules enclosed contained in any arbitrarily selected volume, the proportionally larger are the fluctuations of density.

Usually, in ordinary laboratory experiments, we cannot detect the effect of density fluctuations in gases or liquids, because we deal with large numbers of molecules, and most instruments cannot respond to the fluctuations. However, the blueness of the sky and the critical opalescence of fluids do detect them. In addition, other quantities can also fluctuate. For example, the movement of a small particle undergoing Brownian motion in a liquid (section 4.4.2) is determined by the mean momenta of the molecules within a small volume of liquid, comparable with the volume of the particle—and its jerky movement shows that this quantity fluctuates. In other words, the Brownian motion detects the fact that though the *mean* momentum crossing any plane in the interior of a stationary fluid averages out to zero over a long time, it departs from zero at any instant; it fluctuates about the value of zero. In an analogous sort of way, the pressure exerted by a gas on a very small area of wall fluctuates about its mean value and this can also be detected—not using ordinary sluggish pressure gauges but rapidly-responding ones.

7.7.3 Fluctuations of volume in an elastic system

We can most simply calculate the probability of finding a certain density near a given point in any fluid by fixing attention on a *given number* of molecules and finding *what volume* they occupy. Consider therefore a fixed point X and select the n molecules which at any instant are to be found nearest to it. On the average they must occupy a volume v_0 such that n/v_0 is equal to the mean number-density averaged over the whole volume. But at any instant they may occupy a volume v, greater or less than v_0; of course, the number-density n/v is less probable than n/v_0.*

We may calculate the difference in probability of these two states by calculating the difference of energy between them and then using the Boltzmann factor.

Consider therefore a volume v_0 of fluid in equilibrium at a certain pressure and alter its volume to v by expanding or contracting it, keeping the temperature constant. Following an argument similar to that in 3.5.1, we define

$$\text{(isothermal) bulk modulus } K = -V\left(\frac{\partial P}{\partial V}\right)_T \qquad (3.10)$$

$$= -v_0\left(\frac{\partial P}{\partial v}\right)_T$$

nearly, if the change of volume is not too great. Therefore when the volume has increased from v_0 to v the extra pressure acting is

$$P' = -K\frac{(v-v_0)}{v_0}.$$

The energy required to increase the volume further by dv is

$$-P'\,dv = +K\frac{(v-v_0)}{v_0}\,dv$$

so that the total energy required is

$$\Delta\bar{E} = \frac{K}{v_0}\int_{v_0}^{v}(v-v_0)\,dv = \frac{1}{2}K\frac{(v-v_0)^2}{v_0} = \frac{1}{2}Kv_0s^2 \qquad (7.24)$$

where s is the fractional change of volume, $(v-v_0)/v_0$, Eq. (3.24). This means that the energy is increased whenever it deviates from its equilibrium value, because the squared term must be positive. This is reasonable: the fluid pressure resists any change from the equilibrium value.

The ratio

$$\frac{\text{probability of volume } v}{\text{probability of volume } v_0} = e^{-\Delta\bar{E}/kT} = e^{-(Kv_0/2kT)s^2} \qquad (7.25)$$

and this gives the probability of a volume fluctuation of magnitude s.

---

* It does not matter if any one of the original n molecules diffuses away from the vicinity of X and its place is taken by another because (according to both classical and quantum mechanics but contrary to intuition) molecules are indistinguishable from one another. We are justified in thinking that v_0 or v is always occupied by the same molecules.

We could proceed to work out the mean-square volume fluctuation $\overline{s^2}$ from the correctly normalized probability by averaging s^2. But we can avoid all this by noting that the energy $\Delta \bar{E}$ is a squared term of exactly the same form as the $\frac{1}{2}I\omega^2$ or $\frac{1}{2}mv_x^2$ terms in the kinetic energy of a single molecule. Using the same terminology as before (section 5.3), the compressibility of a volume v_0 of any substance confers on it *one degree of freedom*. However large or small v_0 is, however many molecules it contains, its fluctuation of volume confers one single degree of freedom.

Therefore the mean-square volume fluctuation is given by

$$\tfrac{1}{2}Kv_0\overline{s^2} = \tfrac{1}{2}kT$$

that is

$$\overline{s^2} = \overline{\left(\frac{v_0 - v}{v_0}\right)^2} = \frac{kT}{Kv_0}. \tag{7.26}$$

It is obvious that if v_0 is small, the r.m.s. value of s is large and this agrees with the statement that fluctuations are largest in small volumes.

Eq. (7.26) says that the smaller the bulk modulus K, the greater the fluctuations. The reason is simply that little energy is then required to cause a change of volume so that large changes of volume become probable. Now, the isothermal bulk modulus K is proportional to the slope of the PV isotherm and at the critical point this is zero. Eq. (7.26) could then indicate infinite fluctuations. Actually these cannot occur because higher derivatives of the slope have to be taken into account—but the fact remains that fluctuations of volume (or density) are large and light is strongly scattered. As explained above, the analysis of the optical effects will not be taken further.

It is worth noting that results analogous to Eq. (7.26) for other elastic systems can be written down at once. For example, the length of a rod fluctuates because the atoms in it are in motion. If Y is Young's modulus, its increase of potential energy

$$\Delta \bar{E} = \tfrac{1}{2}Yl_0s^2$$

where l_0 is its mean length, s the fractional change of length $\Delta l/l_0$. Hence

$$\overline{s^2} = \frac{kT}{Yl_0}. \tag{7.27}$$

For a rod 1 m long with $Y = 10^{11}$ dyn/cm^2, the r.m.s. fluctuation of length is of order 10^{-11} cm at room temperature, which conforms with ordinary experience by being negligible.

7.7.4 Fluctuation in a perfect gas

An important result will be derived as a particular case of Eq. (7.26). Consider the fluctuations of a volume containing n molecules of a perfect gas. Since it obeys $PV = RT$,

$$K = -V\left(\frac{\partial P}{\partial V}\right)_T = P \tag{7.28}$$

—the isothermal bulk modulus is equal to the pressure. Thus

$$\overline{s^2} = \frac{kT}{Pv_0}$$

where v_0 is the mean value of the volume. Now $n/v_0 = N/V_0$, where N is Avogadro's number and V_0 is the molar volume at pressure P and temperature T. This gives

$$s = \frac{1}{\sqrt{n}}. \tag{7.29}$$

(This value of s is the r.m.s. fluctuation.) This means that n molecules are contained in a volume which fluctuates between the values

$$v_0\left(1+\frac{1}{\sqrt{n}}\right) \quad \text{and} \quad v_0\left(1-\frac{1}{\sqrt{n}}\right)$$

in the sense that these limits define the r.m.s. fluctuation. We can express this slightly differently by calculating the *number* of molecules contained in a *fixed volume* v_0. If that number on average is n, it fluctuates between probable limits

$$n\left(1+\frac{1}{\sqrt{n}}\right) \quad \text{and} \quad n\left(1-\frac{1}{\sqrt{n}}\right),$$

that is, between $n \pm \sqrt{n}$. For example, a gas under standard conditions contains 3×10^{19} molecules in 1 cm^3 on average; in fact the number has a r.m.s. fluctuation of roughly 5×10^9. Expressed differently, this is a fluctuation of order 1 in 10^{10} which is negligible. This is because the number of molecules is so large. But in a volume of 100 Å cubed, there are only 30 molecules on average, and the number therefore fluctuates between about 25 and 35, which is roughly a 15% variation. Once again, the smaller the number, the relatively larger the fluctuations.

This $\sqrt{n}$ law for a gas of independent particles has been derived in a rather roundabout way. It could have have been deduced more directly from first principles by expressing the fact that the probability of finding a molecule inside v_0 is independent of where all the other molecules are.

Many statistical systems obey this kind of law. The problem of counting the heights of members of a population, referred to at the beginning of section 7.7.2, is typical; numerical examples of the $\sqrt{n}$ rule have been given in section 4.2.

7.8   PROPERTIES OF LIQUIDS ESTIMATED ON VAN DER WAALS' EQUATION

Extending the application of van der Waals' equation even below the critical point into the liquid region is of course quite unwarranted. But it must be remembered that, following the discussion of section 7.2.1, useful laws of corresponding states can be written down. With this in mind we can estimate the tensile strength of a liquid using van der Waals' equation. We will also discuss certain aspects of the boiling of liquids.

7.8.1   Tensile strength and superheating of liquids

We have seen in Fig. 7.10(a) that van der Waals' equation gives isotherms at low temperatures which go below the $P = 0$ axis, corresponding to states of tension. The minimum value of P on any isotherm corresponds to the tensile strength at that temperature.

We have already calculated this value as

$$P = \frac{a(V - 2b)}{V^3} \tag{7.20}$$

(with $V < 3b$ to make sure we are at a minimum). The volume of the condensed phase is roughly equal to b, so the tensile strength at low temperatures should be roughly

$$P = -\frac{a}{b^2} \tag{7.30}$$

which is equal to $-27P_c$, where P_c is the critical pressure. For liquid argon this is badly wrong. We have already quoted (section 7.1.3) that liquid argon has a tensile strength of 12 atmospheres; van der Waals' equation predicts 1,300 atmospheres.

We can interpret this equation by referring back to the relation $a/b \approx \frac{1}{2}L_0$ [Eq. (7.8)], where L_0 is the molar binding energy at low temperatures. Our expression for the tensile strength is therefore equal to the energy required to vaporize a mole of liquid divided by the molar volume. Since pressure or tension is an energy density, we must be implying that when the liquid breaks, it half vaporizes. This is not the mechanism at all, and the van der Waals' approach is unrealistic.

By contrast, van der Waals' equation predicts with surprising accuracy the maximum temperature of superheating, possibly because now the liquid *does* all vaporize. This temperature was called T_m and was defined in section 7.1.2. Referring to the complete family of isotherms, Fig. 7.10(a), we can say that if the branch corresponding to superheated liquid goes below the $P = 0$ axis, then in any experiment at this temperature conducted at low (approximately zero) external pressure, the liquid can be contained. But if the minimum of the isotherm lies above the P-axis, then the liquid cannot exist under low external pressure. T_m is therefore the temperature of that isotherm whose minimum first touches the $P = 0$ axis. This isotherm is labelled T_m in Fig. 7.18. Putting $P = 0$ in Eq. (7.20), the condition is that $V = 2b$, and this in turn gives

$$T_m = \frac{27}{32} T_c. \tag{7.31}$$

In other words, van der Waals' equation predicts that liquids can be superheated, under small external pressures, to 0.85 of their critical temperatures.

In Fig. 7.19, T_m is plotted against T_c for a number of liquids. Data for some of the points are given here:

	T_m	T_c
Sulphur dioxide	323°K	430°K
Ether	416°K	466°K
Alcohol	477°K	516°K
Water	543°K	647°K

The line of gradient 0.85 passes quite well through the points and those for a number of other liquids.

7.8.2 Vapour pressure of liquids

In principle we can calculate the vapour pressure of a liquid once we know where the horizontal part of an isotherm is to be drawn so as to equalize the areas of the loops, as shown in Fig. 7.10(b). Unfortunately it is not possible to do this analytically because the cubic form of the equation makes it impossible to express the integrals in closed form.

However, we may guess that the probability that a molecule can escape into the gas phase will contain the factor $\exp(-L_0/RT)$, and that therefore this factor will appear in the equation for the vapour pressure.

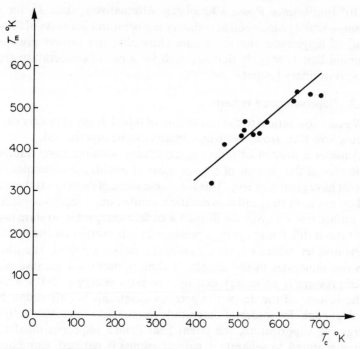

Fig. 7.19. Plot of maximum temperature of superheating T_m against critical temperature T_c for several liquids. Data from Kennick, Gilbert and Wisner, *J. Phys. Chem.* **28**, 1297 (1924).

Thermodynamic and statistical arguments show that a better equation is of the form

$$P = (\text{const.}) \times T^n\, e^{-L_0/RT} \qquad (7.32)$$

where n is equal to the difference of specific heats of vapour and liquid (in units of R), and is therefore a small number. It is quite reasonable that this factor T^n should enter, because the latent heat varies with temperature.

Over small temperature ranges, the exponential factor varies much more rapidly than the T^n term and this allows a quick estimate of L_0 to be made if vapour pressures are roughly known. Comparing vapour pressures at T_1 and T_2, and taking only the exponential term into account,

$$\ln\left(\frac{P_1}{P_2}\right) \approx \frac{L_0}{R}\left(\frac{1}{T_2} - \frac{1}{T_1}\right) = \frac{L_0}{R}\left(\frac{T_1 - T_2}{T_{av}^2}\right) \qquad (7.33)$$

where T_{av} is a mean between T_1 and T_2. Water, for example, has a vapour pressure of about 2 cm at room temperature, that is about 300°K, while at its boiling point 373°K its vapour pressure is atmospheric, 76 cm. Simple arithmetic gives $L_0 \approx 5{,}700\,R$, close to the correct value of

4×10^4 J/mol, since $R = 8.3$ J/mol deg. Alternatively, since L_0 for most common liquids which boil at ordinary temperatures is always of the same order of magnitude, the same data show that the vapour pressure of common liquids roughly doubles itself for a rise of temperature around $10°$ K at ordinary temperatures.

★ ### 7.8.3 Supersaturated vapours

We can now return to the nucleation of liquid droplets in vapours, and discuss how they grow, and why vapours can be supercooled.

Consider a droplet of radius r. Compared with the same number of molecules in the interior of a large mass of liquid, the molecules of the droplet have greater energy. This is because some of them are at the surface and do not have their full coordination number or, in macroscopic terms, the surface tension gives the droplet a surface energy. Or, to state just the same result differently again, a pressure $2\gamma/r$ is exerted on the molecules in the droplet (where r is its radius, γ the surface tension). The distance between molecules in the droplet is slightly decreased because of this. Since pressure is an energy density, the extra energy is $2\gamma V/r$ where V is the volume of the droplet (where, as usual, this is only rough but is good enough for order-of-magnitude estimates). Thus the activation energy for vaporization, the latent heat (which is proportional to the energy required to separate a pair of atoms) is reduced. Consequently the vapour pressure of the droplet, compared with that of a large mass of the liquid with a plane surface, is increased. Instead of a term $\exp(-L_0/RT)$ in the vapour pressure, we have $\exp(-L_0 + 2\gamma V/r)/RT$ where L_0, R and V must all refer to N molecules, that is V must be the molar volume. Thus compared with a mass of liquid with a plane surface ($r = \infty$), the vapour pressure is increased by the factor $\exp(2\gamma V/rRT) = \exp(2\gamma v/rkT)$, where v is the volume of a single molecule, k is Boltzmann's constant, r is the radius of the droplet. It is easy to see why the vapour pressure increases compared with a plane surface. A molecule just inside a curved surface finds it easier to escape because it does not have to break $n/2$ bonds (see section 3.4) but some smaller number, which is reduced all the more as the surface becomes more strongly curved.

Consider now a vapour at pressure P_0 in equilibrium with a liquid with a flat surface. This means, of course, that the liquid is just boiling or the vapour just condensing. The rate of arrival of molecules, $\frac{1}{4}n\bar{c}$/s.cm^2, must be equal to the rate of escape; therefore since n is proportional to the vapour pressure, both rate of arrival and rate of escape must be proportional to the vapour pressure.

Next consider a droplet inside a vapour maintained at a certain pressure P, greater than P_0 so that the vapour would be called supersaturated. Let us calculate the radius of droplet which is in equilibrium. Compared with a plane surface, the rate of evaporation is increased by the factor $\exp(2\gamma v/rkT)$, while the increased vapour density means that the rate of arrival of molecules is increased by the factor P/P_0. Thus for equilibrium

$$\frac{P}{P_0} = e^{2\gamma v/rkT}. \tag{7.34}$$

Thus inside a supersaturated vapour at pressure P, a droplet of radius r_c given by

$$\ln\left(\frac{P}{P_0}\right) = \frac{2\gamma v}{r_c kT} \tag{7.35}$$

is in equilibrium. If (P/P_0) is less than unity, r_c would be negative which means that no droplet is in equilibrium. If $P = P_0$ the surface must be plane. If $(P/P_0) = 1.1$, $\ln(P/P_0) \approx 0.1$ and r_c is of order 10^{-7} cm so that the equilibrium droplet contains the order of 100 atoms.

But we can show that this equilibrium is an *unstable* one. Imagine a vapour held at a pressure P but the radius of the droplet to be slightly *decreased* from its equilibrium value. Then the number of molecules condensing per second remains the same but the number evaporating increases. This tends to make the droplet shrink even further. Conversely, a droplet whose radius slightly exceeds r_c must grow even bigger.

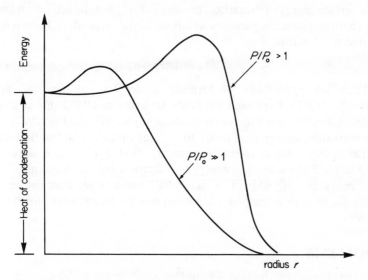

Fig. 7.20. Energy of a droplet of radius r as a function of r. The energy of the same molecules inside a large mass of liquid, with no surface, is taken as zero.

In all the equilibria we have discussed so far, the condition that the forces are zero has been interpreted to mean that the potential energy is a minimum (section 3.2.1). Such a system, displaced, will return to equilibrium. Here however the potential energy of the droplet expressed as a function of its radius must go through a *maximum*. There is certainly equilibrium when the radius has exactly the required value corresponding to the given pressure P, but displaced from this condition (i.e. slightly increased or decreased in radius) the droplet will evaporate or expand further and *not* return to equilibrium.

Figure 7.20 expresses this. Each curve refers to a given degree of supersaturation, a given (P/P_0). It expresses the surprising fact that, at a given temperature with a given pressure of vapour, the molecules condensed as a small droplet can actually have a *higher* energy than when they are in the vapour. The molecules inside the droplet have indeed each lost their energy of condensation (not the full ε per pair but something less because of the finite temperature); but when the droplet is small there are, proportionally, so many molecules at the surface that the droplet as a whole actually has more energy than in the corresponding vapour. Only when the droplet is quite large does the process of condensation reduce the energy of the assembly.

Thus there is an *activation energy* for droplets to form, and this at once explains why vapours must be supersaturated before they can condense. The surface energy of the critical droplet is $4\pi r_c^2 \gamma$ but the activation energy (by thermodynamic arguments which we cannot reproduce here) is equal to one-third of this,

$$A_0 = \tfrac{4}{3}\pi r_c^2 \gamma. \tag{7.36}$$

Thus the probability of forming a droplet contains the factor $\exp(-\tfrac{4}{3}\pi r_c^2 \gamma/kT)$. Expressed explicitly in terms of (P/P_0) this is a complicated function, but it increases dramatically with (P/P_0). For $(P/P_0) = 1.1$, the activation energy is of order 10^{-12} ergs or 1 eV, and the Boltzmann factor is $\exp(-40)$ at room temperature. If (P/P_0) is 1.2, r_c is reduced by a factor 2, the activation energy by a factor 4 and the Boltzmann factor becomes $\exp(-10)$ which is a factor 10^{10} times larger than before. This example, though incomplete, demonstrates the interplay of the different factors.

PROBLEMS

7.1. You are given that the critical temperature of hydrogen is 33°K.
    (a) Write down a relation between the inversion temperature and the critical temperature for a gas obeying van der Waals' equation. What is the inversion temperature of hydrogen?

(b) Show that at room temperature hydrogen gets hotter when it undergoes a Joule–Thomson expansion.

(c) In the 1840's, Regnault found that at room temperature and moderate pressures, PV for hydrogen was greater than RT. He called it a 'more than perfect gas' because all other gases known at the time had PV less than RT. Explain this observation.

7.2. Find the critical constants for Dieterici's equation

$$P(V-b) = RT \exp(-a/RTV)$$

where a and b are constants and all other symbols have their usual meaning.

What is the second virial coefficient and the Boyle temperature for Dieterici's equation?

7.3. Methyl chloride CH_3Cl has a critical temperature of $416°K$. The liquid has density $= 1\,g/cm^3$ at room temperature. The molecular weight is 50.5. The second virial coefficient is

T	239	255	311	366	422	450	$°K$
B	-764	-637	-401	-265	-184	-155	cm^3/mol

Data from J. S. Rowlinson, *Trans. Faraday Soc.* **45**, 974 (1949).

(a) Estimate b and the diameter of a molecule from the molar volume of the liquid.

(b) If van der Waals' equation held, what function of T plotted against B would give a straight line graph? Plot such a graph and though it gives a pronounced curve, strike a reasonable straight line and estimate a. Hence estimate ε.

(c) Estimate ε from the critical data.

(d) The molecule has a Cl^- ion at one end and a concentration of positive charge at the other. It therefore acts like an electric dipole; the dipole moment μ of charges $+e$ and $-e$ separated by distance l is el. In addition to the 6–12 potential energies, two molecules therefore have an additional dipole-dipole interaction. This may be attractive or repulsive depending on relative orientation but on the whole it is attractive of order of magnitude $\mu^2/4\pi\varepsilon_0 r^3$ (J if μ is in Coulomb metres and r is in metres). This swamps the r^{-6} attraction so we are left with a 3–12 potential energy. Sketch the shape of the $\mathscr{V}(r)$ curve. Estimate μ.

(e) Compare it with the moment of charges $\pm e$ (where e is the electronic charge, section 2.1.2) separated by 1 Å. Sketch the charge distribution in the methyl chloride molecule.

7.4. One mole of a van der Waals' gas is kept in a vessel at its critical volume, but its temperature T is greater than T_c. Show that its isothermal bulk modulus is $3R(T-T_c)/4b$. Hence show that the mean square volume fluctuation is given by

$$s^2 = \frac{4}{9N} \cdot \frac{T}{T-T_c}.$$

7.5. A random walker takes N steps in the $+x$ or $-x$ direction (see section 6.3). He takes one step every τ seconds and each is of length l. What is (a) the most probable nett distance travelled? (b) the likely fluctuation from this? (c) Verify the $\sqrt{t}$ law for his diffusion and write down an approximate expression for his diffusion coefficient.

7.6. A very sensitive spring balance consists of a quartz spring suspended from a fixed support. The spring constant is α, i.e. the restoring force of the spring is $-\alpha x$ if the spring is stretched by an amount x. The balance is at a temperature T in a location where the acceleration due to gravity is g.

(a) If a very small object of mass M is suspended from the spring, what is the mean resultant elongation $\bar{x}$ of the spring?

(b) What is the magnitude $((x-\bar{x})^2)^{1/2}$ of the thermal fluctuations of the object about its equilibrium position?

(c) It becomes impracticable to measure the mass of an object when the fluctuations are so large that $\overline{((x-\bar{x})^2)}^{1/2} = \bar{x}$. What is the minimum mass M which can be measured with this balance?

CHAPTER

<h1>8</h1>

Thermal properties of solids

8.1 THE EXTERNAL FORMS OF CRYSTALS

To the physicist, intent on understanding the forces which bind matter together, the ideal form of solid is the crystal. There, the molecules are regularly arranged, so that the environment of any one of them is well defined and the problem of computing the energy of the assembly is comparatively simple. (This enthusiasm for crystalline solids is not shared by engineers, who ask for materials which are mechanically strong. Typically, they use materials which are deliberately made impure and contain more than one phase, and if they are crystalline at all, the crystals are very small.)

The inner regularity of the molecular arrangement in crystals is made manifest by the regularity of their external forms. There is an obvious tendency for crystals to be bounded by plane faces. Any one substance usually forms crystals of a particular *habit* (that is, general shape), whether cubes or needles or hexagonal prisms or flat plates. No two crystals of the same substance are identical in shape; certain facets in one specimen, compared with another, may be enlarged at the expense of others. But the *angles* between corresponding faces are remarkably reproducible.

In Fig. 8.1 for example, one cannot help feeling that the two crystals (*a*) and (*b*) have an underlying identity and that the different relative sizes of the faces is somehow accidental and unimportant. We can express this as follows. We construct *normals* to each face and then let each normal be moved parallel to itself so that it passes through the centre of a sphere. Each normal intersects the sphere in a point and the arrangement of these points on the sphere expresses the angular relationship of the faces.* The two crystals (*a*) and (*b*) give identical patterns, independent of the relative sizes of the faces, as shown in Fig. 8.1(*c*), except of course that by chance certain facets might be absent from one.

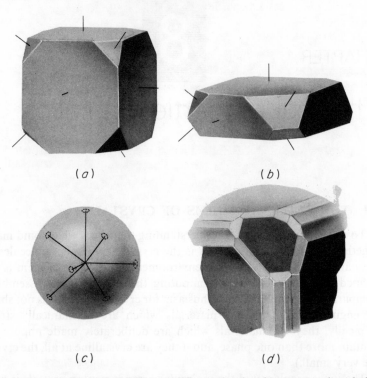

(*a*)   (*b*)

(*c*)   (*d*)

Fig. 8.1. (*a*) and (*b*): Crystals with faces of different sizes but the same angular relationships. Normals to each face are shown. Three cube faces and four octahedral faces are visible. (*c*): When the normals are translated to pass through the centre of a sphere, (*a*) and (*b*) give identical patterns. (*d*): Faces near an octahedral face.

In the very simple crystal forms shown, there are only two different kinds of face—those which are all at 90° to one another and would form a cube if the oblique faces were absent, and the oblique faces themselves which if the cube faces were absent would form an octahedron. We shall refer to these as 'cube faces' and 'octahedral faces'. Faces at other angles

* It is possible to *project* these points on to a plane surface. Such projections are easier to handle than the 3-dimensional spherical patterns.

are possible and indeed frequent, and some which might occur near an octahedral corner are illustrated in Fig. 8.1(*d*). But we shall not pursue this topic because most of the facts about the growth of crystals can be typified by referring only to the simplest types of face.

It is observed that many edges of a crystal may be parallel to one another. Rotation of the crystal about an axis parallel to these edges therefore brings successive faces into parallelism. Such faces constitute a *zone* and the axis is called a *zone axis*.

8.1.1   Optical goniometry

The angles between the faces of any crystal can be measured using a reflecting goniometer. This can be constructed from a spectrometer or any other optical instrument which provides a parallel beam of white light from a collimator, and a telescope focused for parallel light, with a crosswire. Telescope and collimator are placed roughly at right-angles to one another and then clamped and left fixed. At the centre of the system the crystal is mounted, carefully aligned with a zone axis vertical. Each face can act like a tiny mirror, and when a face is set so that it reflects light into the telescope, an image of the slit of the collimator can be seen. If the crystal is turned about a vertical axis so that another face throws up an image in the same way, then the angle turned through by the crystal is equal to the angle between the faces.

The construction of a typical goniometer head, with a crystal mounted on it, is shown in Fig. 8.2. The crystal is stuck on to a metal point so as to be near the optical centre of the system. The head must have sufficient adjustments* to allow the crystal to be accurately aligned. Two horizontal screws at right-angles are necessary to bring it to the centre of the instrument; two concentric circular scales at right-angles are needed to align

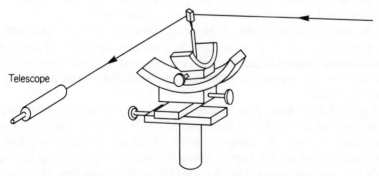

Telescope

Fig. 8.2. A crystal holder for an optical goniometer.

* Instrument makers call them 'degrees of freedom'.

certain edges (and a selected zone axis) of the crystal vertically. The precise adjustments require considerable patience but eventually it is possible to take measurements of the interfacial angles of a whole zone of perhaps a dozen faces. Then the crystal has to be remounted to take measurements on another suitably chosen zone, so that eventually all the angles can be measured and related.

It is advisable to use a white light source, because internal refractions and reflections can give spurious images. With white light these are coloured because of dispersion, and are easily distinguishable from direct reflections from the faces. With a monochromatic source, this would not be so.

With care, measurements accurate to a few minutes of arc can be taken, and the extraordinary symmetry of crystalline form can be revealed, using crystals little bigger than a pinhead. Optical goniometry is one of the most rewarding of practical exercises.

8.1.2   Molecular arrangements; unit cells

We will now show how the faces of a crystal can be related to the arrangement of the molecules inside it. We will in fact study solid argon: other molecular crystals and several metals have similar structures. It will emerge that the crystals are highly symmetrical; cubes and other similar shapes are common. It must be understood that in nature such very high symmetry is not at all frequent. Of all the tens of thousands of crystals which have been catalogued, only about 5% have this high symmetry. Nevertheless, many principles are illustrated by considering these substances, and low symmetries will only be briefly mentioned.

Solid argon crystallizes into a dense lattice (called 'cubic close packed' or 'face centred cubic') where each atom in the interior touches 12 nearest neighbours. Figure 8.3 shows the arrangement; this picture shows a brick-shaped external form with one corner removed.

The most obvious feature of this or any other lattice is its repetitiveness or periodicity. To be precise, there is a unit of repetition of the three-dimensional pattern called the *unit cell*. For ease of illustration however, we will for the moment concentrate on two-dimensional patterns and their unit cells.

Figure 8.4(a) shows a two-dimensional lattice of atoms. *This pattern is not related to Fig. 8.3; it has been chosen only as an example of a two-dimensional lattice.* A regular network, all of whose cells are identical, has been drawn on the lattice. These are unit cells for the lattice. Each contains just one atom. The shape of the cells is such that they can be packed together to cover the area entirely. Further, one unit cell can be translated parallel to one of its edges through a distance equal to the length of that edge and the new position of the atom noted. If the process is repeated parallel to both edge directions, the complete lattice can be built up. (Note that multiples of this unit cell could be used in the same way but one usually chooses the simplest unit.)

Fig. 8.3. Atoms in solid argon (cubic close-packed structure). Some have been removed from a corner so as to form an octahedral face.

There is in fact no unique way of choosing the unit cell for any lattice. Two other possible ones are shown in Figs. 8.4(b) and (c). They each still contain only one atom and indeed each has the same area as that in Fig. 8.4(a). All other possible simple unit cells for this lattice have the same characteristics.

These ideas can readily be extended to three-dimensional lattices. The unit cell may contain several atoms or molecules: for example, the unit

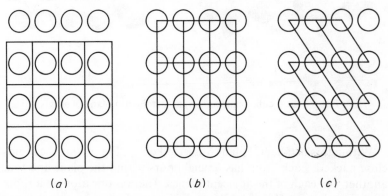

Fig. 8.4 (a), (b) and (c). A two-dimensional lattice, *not related to the argon lattice*, showing possible two-dimensional unit cells.

cell of the argon structure shown in Fig. 8.3 contains 4 atoms. Of the infinity of possible unit cells, one is usually preferred because its symmetry is the same as that of the crystal itself. For the argon crystal, this is a cubic cell with atoms at each vertex and in the centre of each face.

Another obvious feature of the lattice shown in Fig. 8.3, as for any 3-dimensional lattice, is that the atoms are ordered in planes. Further, in this cubic lattice each horizontal plane (for example, the topmost one) is exactly equivalent to the vertical planes which outline the shape and these in fact are parallel to the cube faces of Fig. 8.1(a). Within these planes, the arrangement of atoms is comparatively open, as seen on the left-hand face of Fig. 8.3. One of these cube faces has been dissected out in Fig. 8.5(a). Each atom has 4 neighbours within the plane and it touches another 4 in each of the adjacent parallel planes. If we call the diameter of one atom a_0, then there is one atom inside a square of area a_0^2 in these planes.

In Fig. 8.3, some corner atoms have been removed to leave an oblique octahedral face, exactly corresponding to the octahedral faces of Fig. 8.1. Such a surface would appear flat on the macroscopic scale and this demonstrates how plane faces can develop on crystals and why the angular relations are maintained.

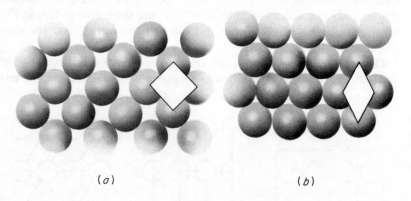

(a)   (b)

Fig. 8.5 (a) A cube plane and (b) an octahedral plane of atoms.

Within this octahedral plane (dissected out in Fig. 8.5(b)), the atoms are close packed. Each atom has 6 neighbours within the plane and touches another 3 in each of the adjacent planes. There is one atom in a rhombus whose angles are 60° and 120° and whose side is a_0—whose area, that is, is $(\sqrt{3}/2)a_0^2 = 0.866a_0^2$.

8.1.3   Surface energies of crystal faces

Let us now calculate the surface energy of a crystal face, in much the same way as we calculated the surface tension of a liquid in section 3.4. Consider first a cube face. Each atom has only 8 neighbours (4 in its plane and 4 in the next plane) instead of 12. To make a new surface we have to cut 4 bonds per atom, each demanding energy $\varepsilon/2$ (energy ε is needed to cut one bond but two surfaces are formed). There are $1/a_0^2$ atoms in 1 cm^2 so that the surface energy of this face is $2\varepsilon/a_0^2$ erg/cm^2. Similarly, the surface energy of an octahedral face is $\sqrt{3}\varepsilon/a_0^2 = 1.73\varepsilon/a_0^2$ erg/cm^2 because each atom has to have three bonds cut and there are $2/\sqrt{3}a_0^2$ atoms/cm^2. The surface energy of the cube faces is therefore *higher* than that of the octahedral faces.

Imagine now a crystal growing with plane faces. Ideally, it ought to grow so as to minimize its surface energy. This means that the octahedral faces should be prominent, the cube faces small. In different words, the cube faces will tend to grow fastest outwards so that they eliminate themselves. This paradoxical statement is illustrated rather sketchily in Fig. 8.6. The shapes are meant to indicate the form of the crystal at successive times. The cube faces grow fastest outwards but get progressively smaller in area, leaving the octahedral faces predominant, as required.

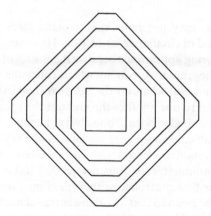

Fig. 8.6. Stages in the growth of a crystal.

In the growth of large crystals, surface energies seem in practice to play little part, though they are important for small ones. In any case, small temperature inhomogeneities which may be caused by the release of energy during the act of crystallization itself, have a profound effect on growth. We will consider some details of the mechanism of growth in section 9.4.1.

Another consequence follows from the different closeness of packing within different planes. Densely packed planes must be further apart (because the total number of atoms in a crystal must be the same, whichever way we count them). Now many crystals can be *cleaved* by being struck sharply in certain directions using a heavy blade. If this is well done, the faces of the two halves are bright and plane. Planes of easy cleavage should—according to these simplified considerations—be those of large spacing, which automatically have small surface energies as we have shown, and are the ones which occur in natural crystals. Frequently this is so.

8.1.4 Single-crystal specimens

Unless they are specially prepared, solid specimens do not usually consist of single crystals, with their atomic planes parallel to one another in all directions. Rather they are polycrystalline with different orientations in different regions so that they can be considered to consist of numbers of small crystals of arbitrary shape all packed together, separated by grain boundaries. Certain properties of solids, notably thermal conductivities at low temperature, vary markedly if grain boundaries are present, and to investigate them it is necessary to convert polycrystalline specimens into single crystals.

A single crystal may not have visible plane faces. It might be, say, in the form of a rod of circular cross-section. However, very often the surface has a shimmering appearance because it consists of little steps parallel to certain crystallographic directions which catch the light, and is not a curved surface at all. But whatever the external shape, the planes of atoms must all be parallel to one another throughout the specimen.

In order to grow single crystals, one always begins with a 'seed' crystal, as perfect as possible. Using this as nucleus, single crystals can be grown provided the *crystallization proceeds very slowly*. In extreme cases, crystals a few millimetres in dimension may take months to form. Temperatures must be controlled within fine limits over these times and all surfaces must be clean so that no other centres of nucleation are present.

If the solid is soluble in water, the seed can be put into a saturated solution from which the water is allowed to evaporate slowly. Molecules crystallize most easily on seeds and the procedure is mainly a matter of common sense—removing other nuclei, keeping the temperature constant and so on.

Growing crystals by cooling the molten liquid is also possible, and is nowadays much more generally used. One method, capable of great variation of detail, is to fill a tube with liquid at a temperature just above the melting point. The tube has a pointed end, and cooling the point causes the liquid to solidify there. In the restricted space, there is probably only room for one nucleus to grow (or else the seed can be put there), so that there is a tendency for a single crystal to form. Subsequently, a carefully controlled temperature gradient is maintained so that the crystal grows along progressively. The alkali halides, like NaCl and LiF, can be grown in this way. They melt at temperatures of the order of 1,000°K. Their vapour pressures are not negligible and the biggest technical problem is to stop them distilling away and depositing on cooler parts of the apparatus. In another temperature range, solid argon rods have been grown by just the same method; this is described in some detail in section 8.5.1.

Another method is 'crystal pulling', where the seed is dipped into the molten liquid and is gradually raised into a cooler part of the apparatus, solidifying and pulling a single crystal rod after it. Finally, in zone melting, the starting point is a polycrystalline rod. By some method of localized heating (such as a narrow beam of radiation, or induction heating), a narrow transverse section is melted. If conditions are right, the surface tension holds the liquid in place so that it does not run away. It is arranged that this melted zone can pass along the crystal, from one end to the other. If the seed is at the starting end, the whole rod eventually becomes crystalline.

8.2   X-RAY STRUCTURE ANALYSIS

The atomic arrangements which have been described have all been elucidated by X-ray diffraction experiments. When a beam of radiation falls on a periodic structure, a grating or a crystal, it is diffracted through large angles if the wavelength Λ is comparable with the repetition length d of the structure. Thus with crystals we are limited to waves whose Λ is a few Ångström units. Any type of waves can be used for diffraction experiments; electrons or other particles of the same wavelength would act in the same way. But there are experimental complications when charged particles are used, and also with neutrons which have a magnetic moment. In any case, the principles governing the diffraction are exactly the same for all types of waves so we will concentrate on X-rays.

When a parallel beam of X-rays of given wavelength falls on a small crystal, most of it travels straight through, but some of the energy is diffracted into a number of beams (Fig. 8.7(a)). These can be recorded on film, or detected by counters and their intensities and angular distribution measured. The scattering is done by the electrons in the atoms of the lattice (the nuclei are so heavy that they are not affected by the X-rays and do not scatter them). The X-rays therefore 'see' spheres whose size is comparable with their own wavelength, and the scattering power of an atom therefore depends on its size as well as the number of electrons in it.

All the diffracted beams from a given crystal are part of the one family. The *angles* at which the beams occur depend on the periodicity of the lattice, that is, on the *size* of the unit cell, while the *intensities* of the beams depend on the *arrangement within* the unit cell. The full family of diffracted beams, expressed as an appropriate function of angles, is called the Fourier transform of the electron density in the crystal.

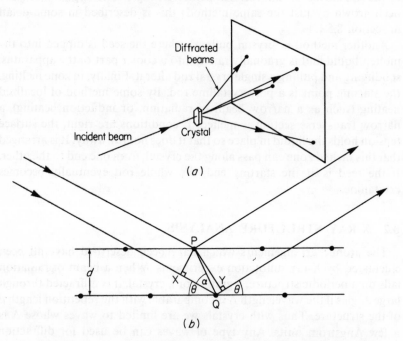

Fig. 8.7. (*a*) A beam of X-rays (covering a wide range of wavelengths) incident on a crystal emerges as a family of diffracted beams. (*b*) Bragg reflection from a plane.

Although it is wrong in principle to take any one diffracted ray and to treat it in isolation from the rest of the family, nevertheless this was done in the early days of X-ray crystallography and has persisted because it is useful in interpreting diffraction patterns. As the method gives quantitative results about *angles* of diffraction using only simple mathematics, we will give the main result.

We isolate a set of parallel *planes of atoms*. In general, these are inclined at some angle to the cube faces but, within the planes, the centres of the atoms lie on some regular pattern, like those of Fig. 8.5. There are many directions in a plane which pass through the centres of atoms, and they do so at regular intervals. Two such lines of atoms (drawn for simplicity as dots) are shown in Fig. 8.7(b), an upper and a lower plane, separated by a distance d. The spacing along the lines is not equal to the distance between planes, and the atoms in the lower plane are not necessarily exactly below the upper ones but are displaced to one side. The third row is similarly disposed with respect to the second, and so on.

Imagine now a beam of X-rays of wavelength Λ incident on an infinite plane of atoms. It can be shown that it diffracts radiation with a strong maximum of intensity in the direction of specular reflection, so that the glancing angle of reflection θ is equal to the glancing angle of incidence, just as if it were a *continuous* reflecting plane. We will only consider the intensity in this direction, and we can speak of the X-rays being reflected from the planes of atoms.

Let us consider interference between radiation scattered by the top plane of atoms and the next plane. The atoms P and Q of Fig. 8.7(b) are typical and we know that if X-rays scattered from P reinforce those scattered from Q (because the phase relation between them is correct) then all other similar pairs of atoms throughout all the planes will also reinforce and the reflected wave will be of finite intensity.

The condition for constructive interference is

$$\text{path difference } XQ + QY = n\Lambda$$

where n is an integer, 1, 2, etc. We must therefore calculate the distances XQ and QY. Call the angle PQX equal to α. Then angle PQY is equal to $(180° - \alpha - 2\theta)$. We have (from $\triangle$PXQ)

$$XQ = PQ \cos \alpha$$

and (from $\triangle$PYQ)

$$YQ = PQ \cos(180° - \alpha - 2\theta)$$

$$= -PQ \cos(2\theta + \alpha).$$

Therefore

$$XQ + YQ = PQ\{\cos \alpha - \cos(2\theta + \alpha)\}$$

which by a trigonometrical identity is equal to $2PQ \sin \theta \sin(\theta + \alpha)$. If we drop a perpendicular from P to the lower plane, we also have

$$d = PQ \sin(\theta + \alpha).$$

Hence the condition for the presence of a diffracted beam is

$$2d \sin \theta = n\Lambda. \qquad (8.1)$$

Notice that θ is the glancing angle with respect to the plane, but neither the spacing within the plane nor the lateral displacement of one set of atoms with respect to the other enters the expression; only the distance d *between parallel planes* enters.

Whereas a *single* plane of atoms reflects X-rays of *any* wavelength specularly, a *stack* of atomic planes only reflects a finite intensity at this angle if, in addition, the condition (8.1) is satisfied. Other wavelengths are cancelled out by interference. Eq. (8.1) is called the Bragg law.

We can now take a single spot of the X-ray diffraction pattern and regard it as having been produced by a Bragg reflection from a stack of parallel planes of atoms, and this enables us to find the spacing d in terms of the wavelength Λ *. For example, with X-rays of wavelength 1.541 Å (produced from a copper target) one particular reflection from an argon crystal was observed to be produced at a glancing angle of 16.52°. Thus

$$2 \times d \times \sin(16.52°) = n \times 1.541 \text{ Å},$$

whence $d = 2.71$ Å if $n = 1$, 5.42 Å if $n = 2$, and so on. By itself, a single reflection does not allow n to be decided, nor is it possible to say which plane did the reflecting. These can only be decided by correlating all the reflections from the whole pattern—intensities as well as angles—and working out just what the unit cell is. (With substances which produce well-shaped crystals, the process is simplified if the reflections can be correlated with the external form.) A check on the value of n comes from comparing the atomic diameter deduced from other methods with that deduced from the unit cell; an error of a factor of 2 is easily detected.

In this way it can be decided that $n = 1$ for this reflection. Thus the side of the unit cell, from Fig. 8.5(a) twice the distance between neighbouring planes, is 5.42 Å, and the diameter of an argon atom is $5.42/\sqrt{2} = 3.83$ Å. Having used other estimates of the diameter to decide which multiple of this figure to use, the X-ray measurement is by far the most accurate.

8.3. AMPLITUDE OF ATOMIC VIBRATIONS IN SOLIDS

Consider a system in a parabolic potential well, whose equation (referred to axes through an origin at the bottom of the well) is $\mathcal{V} = \frac{1}{2}\alpha x^2$. If the system is given energy E it oscillates with an amplitude x_0 given by

---

* X-ray wavelengths were originally measured in absolute units using ruled diffraction gratings. Though comparatively coarse in spacing, they were held at very small glancing angles of incidence, so that the diffraction occurred.

$E = \frac{1}{2}\alpha x_0^2$. A geometrical construction for finding x_0 is shown in Fig. 8.8. A line drawn at height E above the minimum intercepts the parabola at $\pm x_0$, because at $x = x_0$ all the energy is potential energy.

Now consider an atom inside a solid. We will assume that it has $\mathfrak{n}$ nearest neighbours, bound to it by a potential of the 6–12 type. We have already seen (when we estimated the Einstein frequency of vibration in section 3.6.1, Eq. (3.20)) that the atom 'sees' a potential well, due to all its neighbours, which is nearly parabolic and has a curvature near the minimum given approximately by $24\mathfrak{n}\varepsilon/a_0^2$, where a_0 is the separation between atoms.

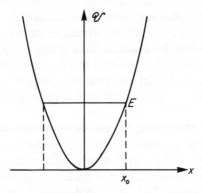

Fig. 8.8. Amplitude of vibration with
energy E in a parabolic potential well
$\mathscr{V}(x) = \frac{1}{2}\alpha x^2$.

If we now extrapolate this assuming that the well is accurately parabolic, its equation referred to axes through the minimum is

$$\mathscr{V}(x) = \frac{12\mathfrak{n}\varepsilon}{a_0^2} x^2 \tag{8.2}$$

where x is the displacement from the minimum. (This result follows from the fact that the parabola $\frac{1}{2}\alpha x^2$ has curvature α at the minimum.)

With these preliminaries, we can now estimate the greatest amplitude of vibration which the atom in a solid can have under ordinary conditions —namely, at the melting point. When the solid is at absolute zero, the atom is (according to the ideas of classical physics) at the bottom of the well. When it is at temperature T and the atom has a mean energy kT, it oscillates within the well—though not, of course, with a strict periodicity because it interchanges energy all the time with its surroundings. Now we know that the melting temperature (at ordinary pressures) is not very different from the temperature of the triple point—and this in turn is always roughly half the critical temperature (80°K and 150°K for argon). We already know that the mean thermal energy of an atom at its critical temperature is about ε, so that at the melting point it is about $\frac{1}{2}\varepsilon$. Thus the amplitude is given by

$$12n\varepsilon\left(\frac{x_0}{a_0}\right)^2 \approx \frac{1}{2}\varepsilon \tag{8.3}$$

that is, $(x_0/a_0) \approx 0.07$, if $n = 10$. Thus we can say that in any substance with a 6–12 potential, the amplitude at the melting point is less than 1/10 of the interatomic separation, which is quite small.

For metals, the ratio of triple point temperature to ε/k is nearer $\frac{1}{4}$ or $\frac{1}{5}$ than $\frac{1}{2}$. For potassium, for example, the melting point is 335°K, and from the data of section 3.7, ε/k is about 0.16 eV so that the ratio is about 0.2. For mercury using data from the same table and a melting point of 234°K, the ratio is about 0.15. But the amplitude at melting is not greatly different from before.

For ionic solids, the calculation can be followed through by noting that an ion 'sees' a potential whose curvature is about $4n\varepsilon/a_0^2$ if the neighbours are spherically disposed about it, as we assumed in the discussion of the Einstein frequency, sections 3.6.1 and 3.8.2. For sodium chloride, $\varepsilon \sim 9$ eV from the data of section 3.8.1, the coordination number is 6, the melting point is 1,073°K. Hence the amplitude at melting is one-thirtieth of the interatomic spacing. Looked at from this point of view, NaCl melts more easily than molecular solids or metals. This is caused by the presence of two sizes of ion in the lattice—small positive ions—and negative ions of about twice their radius. The small ions can slip easily between the large ones, more easily than in lattices where all the members are the same size. Thus though ionic solids like sodium chloride melt at high absolute temperatures, these are really quite low on this proportional scale.

We return to the onset of melting and to some premelting phenomena, considered from a rather different point of view, in section 9.5.2.

8.4   THERMAL EXPANSION AND ANHARMONICITY

It is well known that solids expand when heated. The coefficient of linear thermal expansion is defined by

$$\alpha = \frac{1}{l_0}\left(\frac{\partial l}{\partial T}\right)_P \tag{8.4}$$

where the subscript P denotes an expansion at constant pressure, usually atmospheric pressure. The coefficient of volume expansion β is defined by a similar equation with volume V_0 in place of linear dimension l_0 and since $V_0 = l_0^3$ it follows that $\beta = 3\alpha$. Typically α is of order 10^{-5} per degree for hard solids, 10^{-3} per degree for soft ones, at ordinary temperatures. We will now relate these quantities to the $\mathscr{V}(r)$ curve.

It must be realised at once that if the forces binding one atom to another were purely harmonic—if the potential well between a pair of atoms were exactly parabolic, even for large amplitudes of vibration—then the mean separation of two atoms would always be the same, whatever the amplitude, Fig. 8.9(a). (In the same way, the *mean position* of a simple pendulum remains fixed, whatever its amplitude of swing.) Thus, a solid bound by purely harmonic interatomic forces would not expand with temperature.

The origin of the finite expansion coefficient must lie in the asymmetry of the $\mathscr{V}(r)$ curve, which expresses the fact that two atoms can more easily be pulled apart from one another than pushed together. We can see this graphically, Fig. 8.9(b), by drawing horizontal lines across the well, representing different mean energies and hence different temperatures. With increasing energy, the mean separation tends towards greater separation, and the solid must expand.

Systems in non-parabolic potential wells are said to execute anharmonic motion.

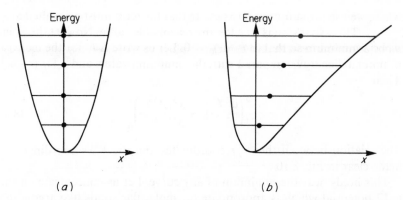

Fig. 8.9. (a) The mean displacement of a simple harmonic motion is always zero, whatever the amplitude. (b) With an unsymmetrical potential energy curve the mean displacement increases with amplitude.

8.4.1   Thermal expansion coefficient

To calculate the magnitude of α, the thermal expansion coefficient, we will first find a convenient approximate equation for the interatomic potential energy curve near the minimum, referred to axes through the minimum. If a_0 is the value of r at the minimum and therefore the separation of a pair of atoms at $T = 0$, we will denote $(r - a_0)$ by x, the displacement from the minimum. Then we will find the limits of x between which a pair of atoms vibrate when their energy is given, and since the positive swing is greater than the negative one, the mean value $\bar{x}$ is greater than zero. $\bar{x}/a_0$ is the total linear expansion between *any pair* of atoms at temperature T, and differentiating with respect to T gives the linear expansion coefficient. This procedure is not exact; it does not properly take into account the effect of the atoms at the *side* of any pair, nor is the mid-point of the swing an exact measure of the mean position; but it is not far wrong.

First we will seek an equation for the interatomic potential energy near the minimum—something better than the parabolic approximation which we used previously (page 45). The anharmonic terms, the extra ones not proportional to x^2, will be the ones responsible for the thermal expansion.

We start by quoting Taylor's theorem. If we have a function $\mathscr{V}$ whose value at $r = a_0$ is $\mathscr{V}(a_0)$, then the value at some other point is

$$\mathscr{V}(r) = \mathscr{V}(a_0) + (r - a_0)\left(\frac{d\mathscr{V}}{dr}\right)_0 + \frac{1}{2!}(r - a_0)^2\left(\frac{d^2\mathscr{V}}{dr^2}\right)_0$$

$$+ \frac{1}{3!}(r - a_0)^3\left(\frac{d^3\mathscr{V}}{dr^3}\right)_0 + \cdots \tag{8.5}$$

where we use the subscript 0 to denote that the term must be evaluated at $r = a_0$. This expansion is true for any reasonable curve. Now let the point a_0 be a minimum so that $(d\mathscr{V}/dr)_0 = 0$. Let us write $\Delta\mathscr{V}$ for the increase of potential energy compared with the minimum value, and x for $(r - a_0)$. Then

$$\Delta\mathscr{V} = \frac{x^2}{2!}\left(\frac{d^2\mathscr{V}}{dr^2}\right)_0 + \frac{x^3}{3!}\left(\frac{d^3\mathscr{V}}{dr^3}\right)_0 + \cdots \tag{8.6}$$

The relation between the $\mathscr{V}, r$ coordinates and the $\mathscr{V}, x$ coordinates is made clear in Fig. 8.10.

This holds near the minimum of any curve. Let us concentrate on the 6–12 potential which is appropriate for molecular solids like argon and with less accuracy (see section 3.7) for metals. Starting with Eq. (3.4):

$$\mathscr{V} = 12\varepsilon\left[\frac{1}{12}\left(\frac{a_0}{r}\right)^{12} - \frac{1}{6}\left(\frac{a_0}{r}\right)^{6}\right] \tag{3.4}$$

$$\frac{d\mathscr{V}}{dr} = -\frac{12\varepsilon}{a_0}\left[\left(\frac{a_0}{r}\right)^{13} - \left(\frac{a_0}{r}\right)^{7}\right] = 0 \quad \text{when } r = a_0$$

$$\frac{d^2\mathscr{V}}{dr^2} = \frac{12\varepsilon}{a_0^2}\left[13\left(\frac{a_0}{r}\right)^{14} - 7\left(\frac{a_0}{r}\right)^8\right] = \frac{72\varepsilon}{a_0^2} \quad \text{when } r = a_0 \quad (3.15)$$

$$\frac{d^3\mathscr{V}}{dr^3} = -\frac{12\varepsilon}{a_0^3}\left[13\times14\left(\frac{a_0}{r}\right)^{15} - 7\times8\left(\frac{a_0}{r}\right)^9\right] = -\frac{1,512\varepsilon}{a_0^3} \quad \text{when } r = a_0.$$

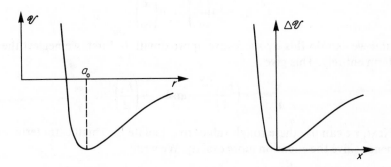

Fig. 8.10. Change of coordinates to an origin through the minimum.

Thus, using the minimum as origin, as in Fig. 8.10, the interatomic potential energy is given by

$$\Delta\mathscr{V} = 36\varepsilon\left(\frac{x}{a_0}\right)^2 - 252\varepsilon\left(\frac{x}{a_0}\right)^3. \quad (8.7)$$

Since we have neglected higher terms and the coefficients are increasing, this approximation is only good up to about $(x/a_0) = 0.1$. But, as we have seen, this is adequate even up to the melting point. The effect of the anharmonic x^3 term is clearly seen in Fig. 8.11.

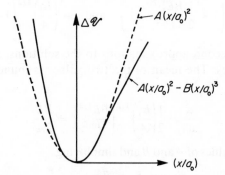

Fig. 8.11. A curve of the type $A(x/a_0)^2 - B(x/a_0)^3$,
for small (x/a_0).

For simplicity, we will write this equation

$$\Delta \mathscr{V} = A\left(\frac{x}{a_0}\right)^2 - B\left(\frac{x}{a_0}\right)^3.$$

Let us now choose a value of the total (kinetic plus potential) energy E_1. We find the limits of swing in the usual way, solving

$$E_1 = A\left(\frac{x}{a_0}\right)^2 - B\left(\frac{x}{a_0}\right)^3,$$

and we can do this by successive approximation. First, we neglect the B term entirely. This gives

$$\frac{x}{a_0} = +\left(\frac{E_1}{A}\right)^{1/2} \quad \text{and} \quad -\left(\frac{E_1}{A}\right)^{1/2}.$$

Next, we can use these rough values to calculate the small extra term and hence solve the equation more exactly. We write

$$E_1 = A\left(\frac{x}{a_0}\right)^2 - B\left(\frac{x}{a_0}\right)^2\left(\frac{x}{a_0}\right) = \left(\frac{x}{a_0}\right)^2\left[A - B\left(\frac{x}{a_0}\right)\right]$$

so that approximately

$$\left(\frac{x}{a_0}\right)^2 = \frac{E_1}{A \mp B(E_1/A)^{1/2}}$$

where the minus sign is associated with the positive square root and vice versa. Taking the square root:

$$\left(\frac{x}{a_0}\right) = \pm\left(\frac{E_1}{A \mp B(E_1/A)^{1/2}}\right)^{1/2} = \pm\left(\frac{E_1}{A}\right)^{1/2}\left(1 \mp \frac{B}{A}\left(\frac{E_1}{A}\right)^{1/2}\right)^{-1/2}$$

$$= \left(\frac{E_1}{A}\right)^{1/2}\left(1 + \frac{1}{2}\frac{BE_1^{1/2}}{A^{3/2}} + \cdots\right) \quad \text{and} \quad -\left(\frac{E_1}{A}\right)^{1/2}\left(1 - \frac{1}{2}\frac{BE_1^{1/2}}{A^{3/2}} + \cdots\right).$$

These are the second approximations to the solutions, good enough for present purposes. The mean of the two values is found by adding and dividing by 2:

$$\frac{\bar{x}}{a_0} = \frac{1}{2}\left(\frac{E_1}{A}\right)^{1/2}\left(\frac{BE^{1/2}}{A^{3/2}}\right) = \frac{BE_1}{2A^2}.$$

Putting in the values of A and B and simplifying:

$$\frac{\bar{x}}{a_0} = \frac{7}{72\varepsilon}E_1. \tag{8.8}$$

This gives the total mean expansion, when the energy of the oscillating atom along the x-axis is E_1. The coefficient of linear expansion is the differential coefficient of this with respect to temperature

$$\alpha = \frac{7}{72\varepsilon} \frac{\mathrm{d}E_1}{\mathrm{d}T}. \tag{8.9}$$

Now we have seen that when the total energy of N atoms (oscillating in 3 dimensions) is $\bar{E}$, the molar specific heat C_p is given by

$$C_p = \frac{\mathrm{d}\bar{E}}{\mathrm{d}T}$$

for any experiment which takes place at constant pressure (including the expansion of a solid). Here, E_1 is the energy of a single atom oscillating in one dimension only, so $E_1 = \bar{E}/3N$. Therefore

$$\alpha = \frac{7}{216\varepsilon} \frac{C_p}{N} \tag{8.10}$$

At high temperatures, *in the equipartition region*, C_p is not very different from $3Nk$ energy units/deg so $\alpha \sim \frac{1}{10} \cdot k/\varepsilon$. For argon, $\varepsilon/k \sim 120$ degrees, whence $\alpha \sim 10^{-3}$ per degree which is the right order for a solid. For metals, for which as we have seen the 6–12 potential can be used for rough estimates, ε/k is of order 3,000 degrees, and $\alpha \sim 10^{-5}$ per degree, again of the right order. Thus the identification of the mechanism causing thermal expansion is correct.

8.4.2   The Grüneisen relation

There is no need to restrict the discussion to high temperatures. The relation (8.10) shows that the expansion coefficient is proportional to the specific heat—which implies that α falls off at low temperatures and this is in fact observed. The expression indeed predicts that α/C_p is a constant for any substance at all temperatures. As it stands, however, the ratio incorporates ε, the interaction energy of a pair of *atoms*, whereas α and C_p are macroscopic properties. It is more convenient to use one of the relations deduced in Chapter 3 to eliminate ε in favour of a macroscopic quantity, and the one usually chosen is the bulk modulus (because it leads eventually to a dimensionless ratio, a pure number);

$$K = \frac{4Nn\varepsilon}{V_0} \tag{3.16}$$

where n is the coordination number and V_0 the molar volume. It must be noted that, strictly speaking, this K refers to adiabatic conditions; we will write it K_{ad}, although previously we did not emphasize the difference between it and the isothermal bulk modulus K_T because we limited the discussion to low temperatures where differences vanish.

Finally, having eliminated ε and having incorporated both a molar volume and an elastic modulus which refers to volume changes, it is reasonable to refer to the volume coefficient of thermal expansion β which is equal to 3α. In these terms our expression is

$$\frac{\beta V_0 K_{ad}}{C_p} = \frac{7}{18} n. \tag{8.11}$$

This ratio should be a constant, independent of temperature, called Grüneisen's constant γ_G for the solid.

We may now invoke a thermodynamic result, namely that the ratio K_{ad}/C_p, that is the adiabatic bulk modulus over the specific heat at constant pressure, is identical with K_T/C_v, the isothermal bulk modulus over the specific heat at constant volume. Thus we can write

$$\gamma_G = \frac{\beta V_0 K_{ad}}{C_p} = \frac{\beta V_0 K_T}{C_v}. \tag{8.12}$$

This is a surprising relation; one would hardly expect the thermal expansion coefficient to be related to the specific heat. In 1908, when Grüneisen first announced his empirical law (in the form α/C_v = constant for any metal) he was unable to account for it. In its complete form, Grüneisen's equation relates changes of pressure, volume and temperature and is often referred to as an *equation of state* for solids.

The absolute value of γ_G which we have calculated is roughly correct. We have already given sufficient data for computing it for solid argon—K_T as its reciprocal the isothermal compressibility in Fig. 3.13(c), C_v and the expansion coefficient in Fig. 5.10(b). Experimentally, therefore, γ_G has the value 2.8 all the way from 20° to about 60°K, falling to about 2.4 at the melting point, 80°K. The simple theory presented here predicts about 4.5 for a close packed crystal structure with $n = 12$.

For metals, the Grüneisen γ_G is usually about 1.4. Ionic crystals usually have γ_G between about 1.5 and 2; the method of calculation for simple ionic crystals is outlined in a problem at the end of the chapter.

8.5 THERMAL CONDUCTION IN SOLIDS

The problem of describing the mechanism of the conduction of heat in solids is one of extreme difficulty. Here we can only try to give a brief sketch of some of the physical phenomena.

We have seen, in sections 6.1.2, 6.1.3 of the chapter on transport processes, that the equations describing the diffusion of molecules and those describing the conduction of heat are formally very similar:

$$\frac{\partial n}{\partial t} = D \frac{\partial^2 n}{\partial x^2} \tag{6.3}$$

and

$$\frac{\partial T}{\partial t} = \left(\frac{\kappa}{C_v \rho}\right) \frac{\partial^2 T}{\partial x^2}. \tag{6.6}$$

Here, n is the concentration of a substance, measured in mol/cm^3, diffusing with time t in a direction x, D is the diffusion coefficient; T is the temperature, $(\kappa/C_v\rho)$ is called the thermal diffusivity. These equations were derived by general arguments and are valid for all states of matter. The symbols represent macroscopic properties of the substance and not the properties of the individual molecules.

The formal analogy between the two equations justifies our using the expression that heat *diffuses* into a body. But we are justified in going further than that. We can look at the process on the atomic scale and bear in mind our success in describing diffusion in gases in terms of the random walk executed by each molecule, in terms of the mean free path. We can therefore try to bring together the concepts of the thermal motion of the molecules of a solid and the ideas of the random walk and the mean free path.

Imagine one small region of a solid to be heated. The additional thermal energy must diffuse to distant regions. Molecules have large amplitudes of vibration there and oscillate violently about their mean positions. Now because of the interactions between molecules in the lattice, this must set their neighbours into oscillation and the result is that a wave disturbance is propagated outwards. It travels with the speed of sound. (The only difference between this kind of disturbance and an audible sound wave travelling through the solid is that the typical frequency, the Einstein frequency, is 10^{10} times greater for the thermal vibrations.)

Now a wave disturbance of this kind can have its direction of energy flow altered—that is, it can be scattered, just as a light beam can be scattered—by a number of processes. For example, it can meet the boundary of the solid when it will be internally reflected or it can meet an imperfection of some kind such as a region of high or low density or an impurity atom of different mass from the bulk. In any case, the energy in the disturbance cannot travel unimpeded through the lattice. Every so often it will have its direction of propagation altered. The average distance that a disturbance travels in a solid before it is scattered is analogous to the average distance travelled by a molecule in a gas, namely the mean free path. We shall denote this average distance in a solid by λ. It is a property of the wave motion and has nothing to do with the distance moved by an individual molecule.

Let us take our analogy between diffusion and thermal conduction one stage further. Let us say that the diffusion of heat is caused by the scattering in random directions, the random walk, of the disturbance carrying the additional thermal energy. Then quoting a result deduced in section 6.5.1, for the diffusion coefficient of a gas of molecules:

$$D = \tfrac{1}{3}\bar{c}\lambda \qquad (6.19b)$$

where $\bar{c}$ is the mean speed of the molecules (practically the speed of sound) and λ is the mean free path. Therefore we expect that for the thermal diffusivity of a solid

$$\frac{\kappa}{C_v\rho} = \tfrac{1}{3}c_s\lambda \qquad (8.13)$$

where c_s is the speed of sound, λ the mean distance travelled by a wave disturbance before being deviated, and the factor $\tfrac{1}{3}$ is perhaps a little arbitrary.

As we have presented it here, we have perforce left the description of the flow of energy rather vague. When it is described precisely, in quantum terms so that the energy can be considered to have particle-like aspects, the analogy becomes exact and the equation for thermal diffusivity can be rigorously justified.

This relation predicts that the thermal conductivity of a solid should be given by

$$\kappa = \tfrac{1}{3}C_v\rho c_s\lambda. \qquad (8.14)$$

In this expression, all the quantities are known—C_v the molar specific heat, ρ the density in mol/cm^3, c_s the speed of sound—except λ the 'mean free path'.

We can therefore compare this equation with experimental measurements and use it to deduce the 'mean free path'; then we can see what the results mean in atomic terms. It will emerge that in a given specimen of material at any given temperature there are *several mechanisms* for scattering the flow of energy and all of these operate at once although with varying effectiveness at different temperatures. Under any given set of conditions, the mechanism which causes the strongest scattering is the one which limits the mean free path and determines the thermal conductivity.

★ **8.5.1   Measurements on solid argon**

The techniques of making specimens of solid argon and of measuring their thermal conductivity are sufficiently unusual to be worth describing. The difficulty is that solid argon exists (under ordinary pressures) only below 83°K, the triple point, which is just above the normal boiling point of liquid nitrogen, 77°K. A rod of solid argon has to be grown, consisting of large crystals if possible (because these turn out to be the most interesting specimens to study); then heat has to be put into one end of the rod and the temperature gradient measured. All these manipulations have to be done inside Dewar vessels at liquid nitrogen temperatures and below, in an argon atmosphere out of contact with air. Then conductivity measurements must be made down to a few degrees absolute, using liquid helium as the refrigerant. The techniques used by Bernè, Boato and de Paz will be described.

The rod was grown inside a pointed glass tube G (Fig. 8.12) as described in section 8.1.4. This tube was surrounded by a copper sheath S, so arranged that its absolute temperature could be accurately controlled while in addition a small temperature gradient could be superimposed. Small electrical heating elements H_1 and H_2 were wound for this purpose on the copper sheath, one at either end, while its lower end dipped into a bath of liquid nitrogen. The level of this bath was kept constant within ± 1 mm, by an automatic topping-up arrangement. The power to H_1 was adjusted to keep the temperature of the lower end of the sheath just *above* the triple point, while H_2 caused the upper end of the sheath to be a few degrees hotter. Thus when pure argon gas was let into the apparatus, it liquefied at the bottom of the glass tube. Next, the power in H_1—electronically controlled—was gradually reduced so that the temperature fell steadily, at about 0.1° per hour. As a result, the temperature at a certain level in the glass tube fell to the triple point and this level travelled gradually upwards—inside the tube the argon solidified up to this level, covered with a thin layer of liquid. The rods, examined after the experiment, were found to consist of comparatively large crystals between 1 mm and 4 mm in size.

Into the top of the tube there hung a long thin plastic rod R with a bead on the end, and the argon crystals grew round this and enclosed it. When the argon rod was long enough, 6 cm long, no more gas was introduced; instead, the surrounding argon atmosphere was slowly pumped away. The argon rod began to evaporate a little from its surface and by gently pulling on the plastic rod (from outside the apparatus) the argon rod could therefore be detached from its glass mould and pulled into the upper part of the apparatus. Next, the liquid nitrogen was pumped away and liquid helium (4°K) syphoned in. At this very low temperature the rod became considerably harder, capable of withstanding the hazards of the next operation.

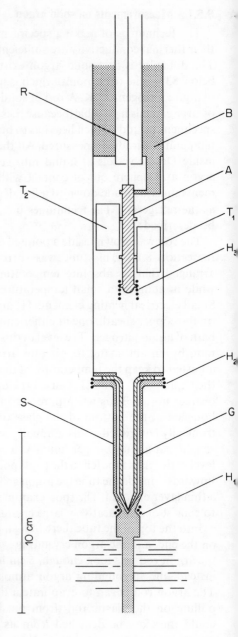

Fig. 8.12. Apparatus for measuring the thermal conductivity of a crystalline solid argon rod A. The rod is shown raised into the upper part of the apparatus, clamped to the copper block B and with its heater H_3 and gas thermometers T_1 and T_2 all attached. G is the pointed glass tube in which the rod was grown, S the copper sheath, extended at the bottom and dipping into liquid refrigerant. H_1 and H_2 are the heaters for controlling the temperature while growing the rod. R is the plastic rod for raising the argon rod out of its mould. The whole apparatus is enclosed in a Dewar vessel. Details have been considerably simplified and no connecting tubes or electrical leads are shown.

Under vacuum the top end of the rod was clamped to a copper block B and a heater H_3 attached to the other end, with two thermometers T_1 and T_2 at points in between. All these attachments were ready in position, with spring-loaded copper clamps held open by nylon strings. The argon rod was slid into position and the strings cut with a blade controlled from outside, so that the clamps closed. If the springs were too strong, the argon rod broke; if they were too loose the thermal contact was poor and the temperatures were measured incorrectly. Of 50 experiments which were started, 12 were carried out to completion, and of these 4 gave results which were self-consistent and judged to be significant.

T_1 and T_2 were helium-filled gas thermometers connected to a sensitive differential pressure gauge, so that the temperature difference (a few tenths of a degree) could be measured directly. The temperature of the copper block at the cold end of the specimen could be varied between $3°K$ and $15°K$, being surrounded either by liquid helium or 'warm' helium gas. The power fed into the specimen was of the order of milliwatts or tens of milliwatts, and the thermometers took several minutes to come to equilibrium.

8.5.2   Impurity scattering of energy flow

The results of measurements of the thermal conductivity of argon on four specimens are given in Fig. 8.13(a). The conductivity is plotted vertically and the temperature horizontally, both on logarithmic scales. At the left of the curves, the lowest measurements were taken near $3°K$; the triple-point, $83°K$, is towards the right. At the bottom of the curves, the lowest conductivity measured, 0.004 watt/cm.deg., is comparable with that of stone or glass at room temperature which are normally regarded as heat insulators. At the top, the best conductivity, 0.6 watt/cm.deg., is better than that of aluminium at room temperature, which is certainly regarded as a good conductor of heat. The 30:1 range of temperature pictured in this graph encompasses a 150:1 range of thermal conductivity and only logarithmic plots can display these wide variations.

Above about $10°K$, the curves are pretty well coincident for all specimens. Therefore, in this 'high-temperature' region, the *dominant* mechanism for scattering the energy flow must be one which does not depend on some variable quality of the specimens (such as their shape or size or purity) but must depend on the bulk properties of solid argon itself. We will return to this region later.

Below $10°K$ however, different specimens have different conductivities, as evidenced by the displacement of the curves parallel to one another. The experimenters in fact noted that the specimens *looked* different from one another. For example, the best-conducting specimen was made from

spectroscopically pure argon and was fairly transparent. The next one
was grown from less pure gas, although it still looked clear; the worst
specimens (actually prepared by a slightly different method from that
described) were 'quite cloudy and opaque'. Since the transparency of a
crystal is an indication of its perfection (see the discussion on the scattering
of light, section 7.7.1), we deduce that the mean free path for diffusion
of the thermal energy is limited, below 10°K, by imperfections in the
crystals. Indeed, we expect wave motions going through regions of
imperfection to be deviated. The imperfections may be grain boundaries,
or small regions of different density caused by strains, or of different
composition caused by impurities.

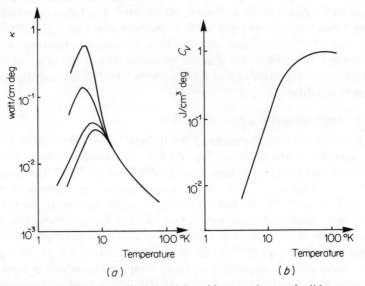

Fig. 8.13. (a) The thermal conductivity of four specimens of solid argon
as a function of temperature, on logarithmic scales. Data from Bernè,
Boato and de Paz, *Nuovo Cimento* **46B**, 182 (1966). (b) Specific heat C_v
plotted on similar logarithmic scales and expressed as specific heat per
cm³. Data from Fig. 5.10(b).

An analogous situation has already been encountered, in section 6.6.1
when discussing thermal insulation with silica powder. In a real gas at
low pressures, the mean free path of the molecules would normally be
large and the thermal conductivity would have a certain value. But if
we fill the space with a fine powder, the mean free path is limited by the
distance between the grains; the conductivity is reduced—and so, presumably, is the diffusion coefficient. In imperfect crystals, the imperfections

(some of which also scatter light) similarly limit the mean free path of the energy flow.

In any given specimen, these imperfections should not change in any way with temperature so that λ should be constant. The density ρ and the speed of sound are also constant; hence the thermal conductivity should be proportional to the specific heat C_v. In Fig. 8.13(b), C_v has been plotted on logarithmic scales—the data of Fig. 5.10 replotted and converted into specific heat per cm^3 using a molar volume of 22.6 cm^3. Now if the conductivity κ is proportional to C_v and both are functions of T, then $\log \kappa$ and $\log C_v$ plotted against T (or $\log T$) should be *parallel* curves. It can be seen that below $10°K$ this is indeed so. This then allows us to measure λ the mean free path. The speed of sound c_s is about 10^5 cm/s. At $5°K$, C_v is equal to 0.0164 J/cm^3.deg. For the best specimen, the thermal conductivity is equal to 0.55 watt/cm.deg, and for worst 0.015 watt/cm.deg. From these data, λ is equal to 10^{-3} cm and 3×10^{-5} cm respectively. These are reasonable figures; for the transparent specimens, λ is many times the wavelength of light, for the cloudy ones it is comparable with the wavelength, as we would expect.

8.5.3   Long mean free paths: the Knudsen region

One's belief in the validity of the energy-flow model is reinforced by the fact that it is possible to prepare crystals (not of argon but of other substances) which are practically free from impurities, so that at low temperatures where the energy density is small, the mean free paths become extremely long, comparable with the dimensions of the crystal itself. The energy wave-trains then behave like a Knudsen gas. This means that when they diffuse down a rod they collide only with the surface of the specimen; there is nothing else to collide with.

Two stringent conditions must be fulfilled before this behaviour can be observed—one concerning the surface of the specimens and the other concerning their physical and chemical purity.

In discussing the molecular flow of gases down tubes, it was not possible to predict whether the molecules stick to the surface and are reemitted at a random angle ('diffuse reflection') or whether they are reflected specularly. Here, we meet an analogous problem. If the surface of the crystal is smooth on the atomic scale, the energy flow will be reflected specularly and the flux of energy down the rod will be unaltered by the collision; on the other hand, if it is desired to make the energy flow undergo a random walk, it must be reflected diffusely and the surface must be rough on the atomic scale.

When a crystal is to be prepared having a long enough mean free path to observe Knudsen-type behaviour, an unusual degree of purity is demanded. This is because the energy flow can be scattered by any departure from perfect regularity in the lattice. Now a foreign atom, of different mass from the rest, can act as a scattering centre. (In the same way, a wave passing down a stretched string is partly reflected at a knot or any small section of different density from the rest.) But any chemical element as found naturally consists of a mixture of isotopes, atoms of different masses. To prepare crystals which are pure enough for the present purpose, the elements therefore have to be pure isotopes.

The most convincing measurements have been made on lithium fluoride, LiF, of which large single-crystal ingots can be grown, using the pointed-tube technique (section 8.1.4). Lithium as it occurs naturally is predominantly ^{7}Li with about 7.5% of ^{6}Li; fluorine is pure ^{19}F. The lighter ^{6}Li was therefore extracted before the LiF was prepared (and a later set of experiments showed that the thermal conductivity was thereby increased). Several specimens were cut from the same big crystal, in the form of square-section rods of different sizes. Finally, the surfaces were sand-blasted in an attempt to make them rough on the atomic scale. This treatment did not so much roughen the surface as produce a thin layer of damaged crystal structure just below the surface and a detailed analysis of the measurements showed that the most of the reflection was diffuse.

The results are shown in Fig. 8.14(*a*), together with specific heat measurements in (*b*), both on log-log scales. Below 20°K, the curves are parallel showing that κ is proportional to C_v, but the conductivity is larger the bigger the cross-section of the specimens. Though the curves look as if they are close together, they in fact show that κ for the 7 mm specimen is between 5 and 10 times that for the 1 mm specimen. Note the enormous value of κ at 20°K—about 200 times better than that of copper at room temperature, an astonishing fact when one considers that LiF is an insulator, since one is used to the fact that they are worse conductors of heat than metals.

The speed of sound is 5×10^5 cm/s. Using all the data, it can be calculated that the mean free path in the 7 mm × 7 mm specimen is about 3 mm; in the 1 mm × 1 mm rod it is about 0.6 mm. Thus the mean free path is certainly of the same order as the dimensions, as predicted by the Knudsen-type theory.

★ **8.5.4   High temperature behaviour**

Above about 10°K for solid argon or 20°K for lithium fluoride, the thermal conductivity *decreases* with rising temperature. At high temperatures in the equipartition region, the graphs on log/log scales tend to become straight lines at 45° showing that $\kappa \propto 1/T$. Since C_v is constant, this implies that the mean free path $\lambda \propto 1/T$. As remarked previously, the fact that impurities or the mode of preparation of the specimens have no effect on this part of the curve—all graphs for different specimens of any substance are coincident—shows that the mean free path is dictated by some property of the lattice itself.

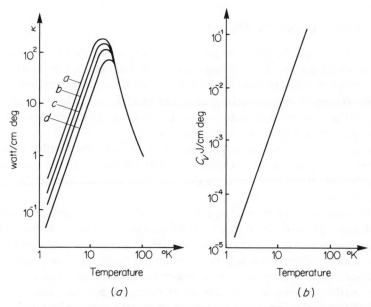

Fig. 8.14. (*a*) The thermal conductivity of four extremely pure LiF rods as a function of temperature on logarithmic scale. The rods were nearly square in cross section of side approximately 7 mm (specimen *a*), 4 mm (specimen *b*), 2 mm (specimen *c*) and 1 mm (specimen *d*). Data from Thacher, *Phys. Rev.* **156**, 975 (1967). (*b*) Specific heat per unit volume below 20°K, the low temperature tail of the curve, far below the equipartition region. Data from Scales, *Phys. Rev.* **112**, 49 (1958).

It is now accepted that this is the anharmonicity of the lattice vibrations. When two wave disturbances pass through one another, they create *new frequencies* which were not originally present in either. This process has no analogy under ordinary conditions for light waves, for example; if it did, then a red beam passing through a green beam might generate ultraviolet and this does not occur. The generation of new frequencies from two wave motions is called 'frequency mixing' and electronic engineers are familiar with related effects using nonlinear circuit elements; because of anharmonicity, a solid is similarly said to be a nonlinear medium to the passage of sound waves. When a wave passes over another and generates a third frequency, this can undergo a sort of Bragg reflection with the nett effect that the direction of energy flow is altered. It is plausible that the effect should become more important as the amplitude of the atomic motions is increased—that is, as the temperature is raised. This is why the thermal conductivity decreases as the temperature increases.

8.6   ELECTRONS IN METALS

Metals can be described as lattices of ions, permeated by a kind of gas of electrons. We have already described (sections 2.1.2, 2.1.4) how some electrons inside atoms are tightly bound, while others are only loosely bound. It is the loosely bound ones which can wander through the lattice. They can be accelerated by electric fields and their energy can be increased by raising the temperature.

In many metals such as copper, silver or sodium, the number of free electrons is about one per atom. In a molar volume of metal, therefore, there are about 6×10^{23} free electrons. We shall call this number '1 mole of electrons'.

The most obvious property of metals is that they are good electrical conductors. But not only do metals conduct electricity, they are also much better conductors of heat than insulators. For example, at room temperature, sodium chloride which is an electrical insulator has a thermal conductivity of 0.06 watt/cm.deg, while metallic copper has $\kappa = 0.92$ watt/cm.deg. Among liquids, alcohol at room temperature and sodium at 200°C have thermal conductivities of 0.0017 and 0.82 watt/cm.deg respectively; again the metal has the much higher thermal conductivity. Further, the thermal conductivity among metals increases roughly in proportion to the electrical conductivity σ. For example, copper and zinc have electrical conductivities of 0.63×10^{-6} and 0.16×10^{-6} (ohm cm)$^{-1}$ respectively, that is in the ratio 3.8:1. Their thermal conductivities are respectively 0.92 and 0.265 watt/cm.deg, which are in the ratio 3.5:1. This rough proportionality holds for all metals. This fact was discovered experimentally by Wiedemann and Franz in 1853, who stated that κ/σ = constant for many metals at room temperature. In 1881, the Danish physicist Lorentz made measurements between 0°C and 100°C and was able to restate the law with another factor present—in the form

$$\frac{\kappa}{\sigma T} = \mathscr{L}, \tag{8.14}$$

a constant for all metals for all temperatures. It is called the Lorentz ratio or the Lorentz constant.

Typical values of κ and σ for a number of metals at 20°C (293°K) are given in the table. The conductivities cover a seven-fold range but the Lorentz ratio agree within 10% of one another—although some metals which are poor conductors do not agree so well.

Metal	σ (ohm cm)$^{-1}$	κ watt/cm.deg	$\kappa/\sigma T$, $T = 293°$K
Tin	0.087×10^6	0.64	2.51×10^{-8}
Zinc	0.174	1.18	2.32
Aluminium	0.354	2.36	2.28
Copper	0.591	3.96	2.28

If we study one metal, copper, over a wide range of temperatures, we find that the Lorentz ratio remains fairly constant within 20% or so from 1,000°K down to 100°K. Below that temperature it changes quite rapidly, however (Fig. 8.15).

From the proportionality of electrical to thermal conductivity over a wide range of temperature, we can surmise that the electrons are responsible not only for the transport of electricity, but also for heat conduction—and that this mode of heat transport is far more effective than other modes of heat transport through the lattice of a metal.

There is one further experimental fact which intimately concerns these electrons in metals. Surprisingly, their specific heat is extremely small. If they behaved like a classical gas, and if 1 mole of metal contained about 1 mole of mobile or conduction electrons, then they would be expected to contribute about $\frac{3}{2}R$ to the measured molar specific heat of the metal. Copper for example would be expected to have a specific heat of $\frac{6}{2}R = 25$ J/mol deg from the lattice and $\frac{3}{2}R = 12.5$ J/mol deg from the electrons, a total of 37.5 J/mol deg. But at ordinary temperatures the observed value is not very different from that expected for the lattice alone so that all that can be said is that the contribution from the electrons is small. At extremely low temperatures however, below about 2°K, the lattice specific heat is expected to be vanishingly small yet the measured specific heat is not small and varies proportionally to T, Fig. 8.16(a). It can be represented by

$$C_v = 7.5 \times 10^{-4}T \text{ J/mol deg}$$

for copper. This is interpreted to be the contribution due to the electrons. If we extrapolate it still using the linear law up to room temperature, it gives only a small contribution of 1% to the total instead of the expected 33% and this at least is self-consistent, (Fig. 8.16(b)). It follows inevitably that electrons in a metal do not behave like a classical gas.

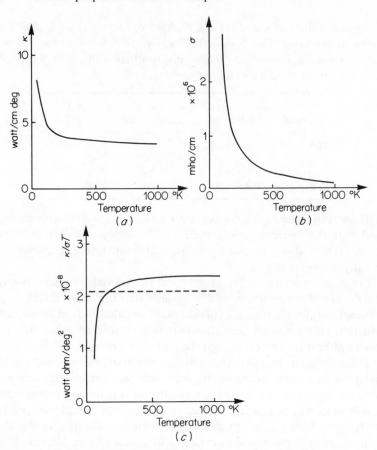

Fig. 8.15. Thermal conductivity, electrical conductivity and the Lorentz ratio of copper as functions of temperature.

So far, we have assembled some experimental facts about metals. The plan for the following section will be to show how some of the properties of the electron gas can be deduced by harmonizing the Wiedemann–Franz law with the unexpected observation about the specific heat. The result will be in the form of an equation for the energy of the electron gas as a function of temperature—which is certainly non-classical in form and can only be understood in terms of quantum mechanics. Historically, the problem was not approached in this way, but the electrons were assumed to behave like a classical gas; it was found possible to explain the electrical and thermal conductivities of metals but their specific heats remained mysterious. Because of its historical interest, we will reproduce this calculation also.

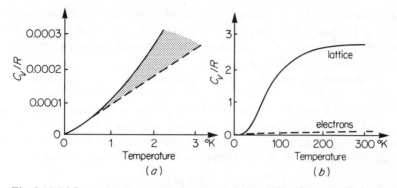

Fig. 8.16. (a) Low temperature measurements, below 2°K, of the specific heat of copper. The predicted contribution from the lattice is shown as a shaded area, the linear part is interpreted to be the contribution from the electrons. (b) Measurements up to room temperature. The lower line is the extrapolated contribution from the electrons.

★ **8.6.1   Thermal and electrical conductivities of metals**

We calculate the thermal conductivity of a metal by imagining the lattice not to exist, but the space occupied by the metal to be filled with the electron 'gas'. This transports heat like an ordinary gas, although with the appropriate values of mean speed and specific heat. The electrons must be presumed to have a finite mean free path. Just what mechanism is responsible for altering the trajectories of the electrons need not be specified; it might be collisions between electrons (a plausible suggestion but not a correct one) or collisions with imperfections or impurities or other departures from perfect regularity in the lattice such as thermal vibrations. Then

$$\kappa = \tfrac{1}{3} C'_v \bar{c} \lambda \tag{8.15}$$

where $\bar{c}$ is the mean speed of the electrons, λ the mean free path and C'_v is the specific heat per cm^3 of electron 'gas'. If there are n_0 electrons/cm^3 inside the metal, then

$$C'_v = \frac{n_0}{N} C_v$$

$$\kappa = \frac{1}{3} \frac{n_0}{N} C_v \bar{c} \lambda$$

where C_v is the specific heat of 1 mole of electrons.

Next we will calculate the electrical conductivity of the same 'gas'. Imagine the metal in the form of a rod or wire of cross-sectional area

A and length l. Let a voltage $\mathbf{V}$ act between the ends. Then the *electric field* is $\mathbf{V}/l$, and if the charge on the electron is e, the force on it is $e\mathbf{V}/l$. This causes the electron to accelerate, with an acceleration $e\mathbf{V}/lm$, where m is the mass of the electron.

If the mean free path is λ and the mean speed is $\bar{c}$, then the mean *time* between collisions is $\lambda/\bar{c}$. We will assume that after a collision, the electron is brought to rest. During its mean free time it accelerates and reaches a final velocity equal to the acceleration multiplied by the time. It is brought to rest by the next collision and then the process repeats itself. Thus the *mean* velocity with which the electron drifts along the wire is

$$\bar{v} = \frac{1}{2}\frac{\mathbf{V}e\lambda}{m\bar{c}l}.$$

If there are n_0 electrons per cm^3, the number drifting across any plane in the wire is $\bar{v}n_0 A$, so the charge transported per second is

$$\text{Current} = \bar{v}n_0 Ae = \frac{1}{2}\frac{\mathbf{V}e^2\lambda n_0 A}{m\bar{c}l}.$$

Ohm's law states

$$\text{current} = (\text{conductivity } \sigma)\frac{A\mathbf{V}}{l}$$

so that we have derived Ohm's law and

$$\sigma = \frac{1}{2}\frac{n_0 e^2 \lambda}{m\bar{c}}. \tag{8.16}$$

Let us compare this expression with that for the thermal conductivity. Both contain the mean free path λ, which is not surprising for transport processes with the same carriers. If we form the ratio, λ cancels out:

$$\frac{\kappa}{\sigma} = \frac{\frac{1}{3}(n_0/N)C_v\bar{c}\lambda}{\frac{1}{2}(n_0 e^2/m\bar{c})\lambda} = \frac{2}{3}\frac{C_v m\bar{c}^2}{Ne^2}. \tag{8.17}$$

Now let us rewrite this expression in slightly more general terms. $\frac{1}{2}m\bar{c}^2$ is the mean kinetic energy of an electron—and we can assume that it is not very different from $\frac{1}{2}m\overline{c^2}$ (notice the different averaging). Let us denote the mean energy of N electrons by $\bar{E}$:

$$\bar{E} = \frac{1}{2}Nm\overline{c^2}$$

The specific heat C_v is equal to $\mathrm{d}\bar{E}/\mathrm{d}T$. Therefore

$$\frac{\kappa}{\sigma} = \frac{4}{3}\frac{\bar{E}\,\mathrm{d}\bar{E}/\mathrm{d}T}{(Ne)^2}.$$

Experimentally, $\kappa/\sigma T = \mathscr{L}$ the Lorentz constant. Hence,

$$\bar{E}\,d\bar{E}/dT = \tfrac{3}{4}(Ne)^2\mathscr{L}T$$

so that

$$\bar{E}^2 = E_0^2 + \tfrac{3}{8}(Ne)^2\mathscr{L}T^2$$

where the term E_0^2 is an arbitrary additive constant. We can take the square root of both sides, making the assumption (which we will justify shortly) that E_0^2 is much larger than the T^2 term under our conditions. Then

$$\bar{E} = E_0\left[1 + \frac{3}{8}\frac{(Ne)^2\mathscr{L}}{E_0}T^2\right]^{1/2}$$

$$= E_0 + \frac{3}{16}\frac{(Ne)^2\mathscr{L}}{E_0}T^2 + \cdots. \qquad (8.18)$$

This is the variation of energy of the electron gas with temperature which we deduce from the observation that the Lorentz ratio is a constant.

Let us see what specific heat it predicts. We write

$$C_v = \frac{d\bar{E}}{dT} = \frac{3}{8}\frac{(Ne)^2\mathscr{L}}{E_0}T.$$

It can be seen at once that at any rate this is of the correct form, in the sense that it agrees with observations at very low temperatures. If we put in numbers, we can find the one unknown E_0.

Putting $C_v = 0.002T$ J/mol deg deduced from measurements at low temperatures, $Ne = \mathscr{F}$ the faraday equal to 10^5 coulombs, the Lorentz constant equal to 2.5×10^{-8} watt ohm/deg^2, then $E_0 = 50{,}000$ J/mol—an enormous value equivalent to 1 eV for each electron. (At least our assumption that $E_0^2 \gg$ the term in T^2 is justified and our method of taking the square root is self-consistent.)

Thus the discussion of the Wiedemann–Franz law and the small specific heat of the electron gas leads us to conclude that *even at absolute zero* the electrons have enormous energies, the E_0 term. Being a constant, E_0 does not show up in the specific heat measurements which only measure $d\bar{E}/dT$: all these detect is the coefficient of the small T^2 term in the energy, and at ordinary temperatures the energy of the electrons changes by very little.

These facts can only be explained in terms of quantum mechanics—in particular the uncertainty relation and the exclusion principle—which we will not attempt to do. All we have achieved, while suggesting a mechanism for the conduction of electricity and heat in metals, is to point out yet another system to which the Maxwell distribution and the law of equipartition of energy are not applicable. For completeness, we will mention that the specific heat of the electron gas as a function of temperature has been calculated quantum-mechanically and is shown in Fig. 8.17. It does eventually reach the equipartition value of $\frac{3}{2}R$ expected of a monatomic gas, but only at the impossibly high temperature of 10^4 °K —when the metal would have ceased to exist.

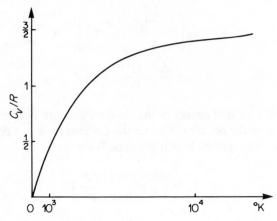

Fig. 8.17. Predicted specific heat of 1 mole of electron gas.

★ **8.6.2 The classical calculation**

We have already explained that the approach which we have just adopted was not taken by the pioneers in the subject. It was at the very end of the nineteenth century that electrons ('atoms of electricity') were discovered in ionized gases and their elementary charge roughly measured by J. J. Thomson. Within a few years, Lorentz and Drude were proposing that electrons formed a gas inside metals, in much the same way as we have done. But it was natural in those days, before the advent of even the most rudimentary form of quantum theory, to assume that the electrons really did behave like an ordinary gas obeying the laws of classical physics.

The ratio of conductivities can be written

$$\frac{\kappa}{\sigma} = \frac{4}{3} \frac{C_v \bar{E}}{(Ne)^2}$$

on either theory. If now we assume the classical equipartition laws:

$$C_v = \tfrac{3}{2} Nk, \qquad \bar{E} = \tfrac{3}{2} NkT, \qquad (5.12)$$

then we find

$$\frac{\kappa}{\sigma T} = 3\left(\frac{R}{\mathscr{F}}\right)^2 = 3\left(\frac{k}{e}\right)^2.$$

This is a very precise prediction—and it almost agrees with experiment. Putting R the gas constant equal to 8.31 J/mol deg and $\mathscr{F}$ the faraday to 10^5 coulomb (or k, Boltzmann's constant, equal to 1.38×10^{-23} J/deg, e, the electron charge, to 1.6×10^{-19} C) the Lorentz number is predicted to be about 2.1×10^{-8} watt ohm/deg^2 which is remarkably good. But of course the specific heat is completely incorrect and it was because of this that the whole theory had eventually to be demolished and rebuilt in quantum terms—a process which took more than 20 years.

PROBLEMS

8.1. An imaginary element of atomic weight 80 has density 1.2 g/cm^3. Its crystal structure is known to be simple cubic and an experiment is carried out to determine its lattice spacing accurately. Using Cu Kα radiation ($\Lambda = 1.54$ Å), a reflection from a cube plane is observed at $\theta = 73°\,45'$. What value of n must be used in Bragg's equation for this reflection and what is the lattice spacing? At what angles θ would reflections from octahedral planes be expected?

8.2. An X-ray reflection from a crystal occurs at $\theta = 45°$ when the crystal is maintained at 0°C. When it is heated to 100°C, θ decreases by 3.42 minutes of arc. What is the expansion coefficient of the substance? How is the argument affected if the crystal is not cubic in symmetry?

8.3. Calculate the expansion coefficient and Grüneisen constant for an ionic crystal. Start with the pair potential of Eq. (3.27).

(a) Using Taylor's theorem, show that the equation near the minimum is

$$\Delta\mathscr{V} = \left(\frac{p-1}{2!}\right)\frac{\alpha e^2}{4\pi\varepsilon_0 a_0}\left(\frac{x}{a_0}\right)^2 - \frac{(p-1)(p+4)}{3!}\frac{\alpha e^2}{4\pi\varepsilon_0 a_0}\left(\frac{x}{a_0}\right)^3 \cdots.$$

(Here α is the Madelung constant, not the expansion coefficient.)

(b) Show that in the equipartition region, the linear expansion coefficient is equal to

$$\left(\frac{p+4}{p-1}\right)\left(\frac{4\pi\varepsilon_0 a_0}{\alpha e^2}\right)k,$$

where k is Boltzmann's constant. Estimate it for KCl for which $a_0 = 3.1$ Å and Madelung constant $= 1.75$.

(c) Use expressions for the compressibility and molar volume taken from section 3.8.2. Hence show that the Grüneisen constant is $(p+4)/9$.

(d) Compare this with the experimental measurements on KCl. (Data from G. K. White, *Phil. Mag.* **6**, 1425 (1961)).

T °K	β deg^{-1}	C_v J/mol. deg.	V_0 cm^3	K_T dyn/cm^2
30	9.90×10^{-6}	8.31	36.7	1.95×10^{11}
65	51.6	28.6	36.75	1.93
283	111.3	48.7	37.4	1.69

CHAPTER

Defects in solids: Liquids as disordered solids

9.1 DEFORMATION OF SOLIDS

In section 3.7.1 we calculated the deformation produced in a solid by the application of pressures or tensions to it. We will now extend the discussion to deformations which are so great that the solid breaks.

The conventional nomenclature is to call the fractional deformation the *strain* (as already mentioned) and the force per unit area the *stress*.

Previously, we considered only hydrostatic stresses, that is pressures or tensions acting uniformly over the whole surface of the body (like those which exist inside liquids, which cannot support shear strains). In these the atoms in the body become uniformly squeezed or separated from their neighbours in all directions equally. Now in practice, one is very often concerned not with these hydrostatic forces but with tensions or pressures acting along a line, or else with twisting forces. However, we have noted (in section 3.5) that the order of magnitude of the bulk modulus—the initial slope of the stress/strain curve—is of the same order of magnitude as the other elastic moduli so we will continue to deal only with uniform hydrostatic stresses.

We showed in section 3.7.1 that the strain s induced by a stress P is given approximately by

$$P = -K(s - \tfrac{9}{2}s^2 + \cdots) \tag{3.25}$$

assuming that higher terms can be neglected for the small values of s we deal with in practice. (A positive stress is a pressure which produces compression, a positive strain is an increase of volume; whence the minus sign in front of the leading term.) If we plot this function up to large values of s, beyond the range shown in Fig. 3.15, P should go through a maximum. In practice, this result can only mean that there is a maximum tension which the solid can withstand, and when this tension is exceeded the solid breaks.

Therefore the *tensile strength* of a material should be given by the condition

$$\frac{dP}{ds} = 0,$$

that is,

$$-K(1 - 9s) = 0$$

$$s_m \approx 10\%$$

where s_m is the tensile strength.

Our simple theory predicts that a body should be able to withstand a 10% strain. Though we have derived this result only for hydrostatic tensions, the same result should hold for stretching or twisting strengths: a wire ought ideally to be able to be stretched by 10% in length before breaking, a rod should be capable of being sheared through $\frac{1}{10}$ radian before shearing off.

9.1.1 Ductile, brittle and plastic materials

The behaviour of real metals is in strong contrast to this result. Some typical stress/strain graphs are shown in Fig. 9.1. In (a), the body is strained by only a small amount—say 0.001% to 0.01%. Then, when the stresses are removed, the body returns more or less exactly to its original shape and size. However, there exists a strain (typically of order 0.01% to 0.1%) called the *elastic limit*, marked E on Fig. 9.1(b); if the body is strained beyond E, and the forces then removed, it does not return to its original shape and size but remains permanently distorted. Subsequent stressings follow a complicated pattern, but when the strain is of the order 0.1% to 1%, the metal breaks. This means that metals are a factor of 10 or 100 times weaker than our simple theory predicts.

Materials which behave in this way are called *ductile*. This refers to their property of distorting permanently before breaking.

Metals can be prepared as single crystals and these are often very soft and ductile. They can be deformed by squeezing in one's fingers and in extreme cases will even flow under their own weight. If pulled, they can be extended to many times their own length—the strain can reach 10 or

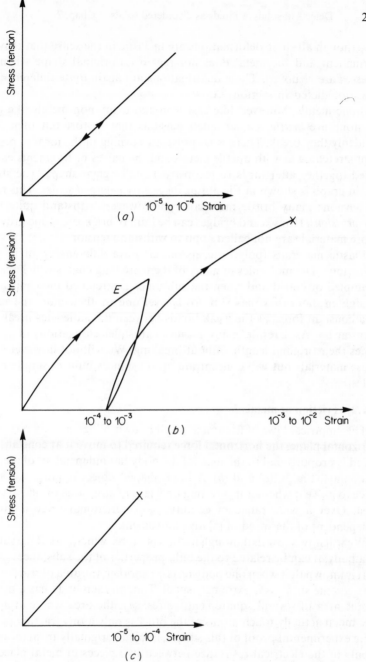

Fig. 9.1. Stress (tension) strain curves for (a) ductile material in the
elastic range, (b) ductile material strained beyond the elastic limit E,
(c) brittle material. The curves are not to scale.

20—though all these deformations are inelastic in the sense that they are permanent and the metal does not regain its original shape when the stresses are removed. These deformations are again quite different from those predicted in section 9.1.

Some metals, however, like cast iron and many non-metals like glass or stone are *brittle*. Under small tensions they distort but then quite suddenly they break. There is no previous warning in the form of permanent stretching as with ductile metals and the halves or fragments can be fitted together afterwards to reconstruct the original shape. The stress/strain graph is shown in Fig. 9.1(c), *on greatly enlarged scales*. This refers to tension: many brittle materials can, however, withstand quite large *compressions*. Houses and bridges can be built of brick and stone provided those materials are not called upon to withstand tension.

Plastic materials (polythene, nylon, etc.) are different again in their behaviour. The molecules of many of them are long chains which may be crumpled or coiled and when the substance is stretched the chains may straighten; later they may slip slowly over one another under the action of a constant force, as the weak bonds between the molecules break and form again. As a result, many plastics can suffer elongations of several times their original length without breaking. We will not however study these materials but will concentrate on substances built of simple molecules.

★ 9.1.2   **Friction of metals**

The empirical laws of friction are well known. When a body rests on a horizontal plane, the horizontal force required to move it at constant low speed is proportional to the mass of the body but independent of the area of contact, Fig. 9.2(a) and (b). A brick-shaped object requires the same force to move it whether it is resting on a face of large area or one of small area. Over a wide range of conditions, the frictional force is almost independent of the speed of relative movement.

We will now show that though friction is a force which acts at the surface of a body, it can be related to the bulk properties of the substance.

It is known that when one body rests on another, the area of real contact between the surfaces is extremely small. Though each body has a macroscopic area of several square centimetres, say, the area where atoms of one metal actually touch atoms of the other is only a tiny fraction of this. (The experimental proof of this statement came originally from measurements of the electrical resistance between two pieces of metal placed in contact under only small forces.) The reason is that no metal surface is plane on the atomic scale. There are always large bumps or asperities even on surfaces which appear to be smooth—at least, if one uses the

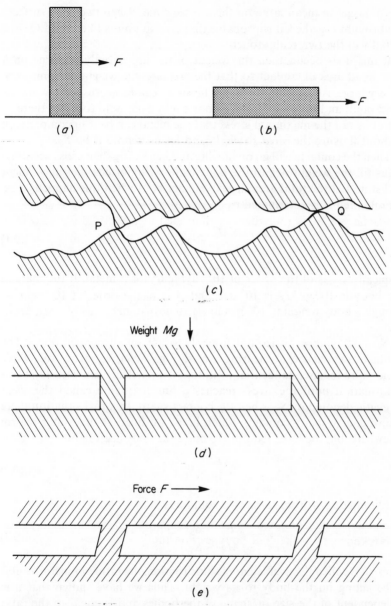

Fig. 9.2. (a), (b) The force required to move a given object at low speeds against friction is independent of the area of contact with the plane. (c) Points of real contact between two bodies. (d) Simplified model of contacts. Metal flows until the pressure due to the weight is equal to the yield strength of the metal. (e) Shearing of contacts by horizontal force.

word 'large' to mean large on the atomic scale. When two such surfaces are brought together, it is probable that only at points like P and Q (Fig. 9.2(c)) will the two really touch.

If this does occur, then the weight of the upper body is borne on a very small area of contact so that the *pressure*, the weight per unit area, is very large. As a result, the metal in these regions is crushed beyond its yield point and it flows. The two bodies may even weld together there.

Let us call the maximum stress that the metal can bear in compression without flowing the *yield strength*, and let us denote it by S_Y.

Then the contacts will go on flowing and increasing their cross-sectional areas till the weight Mg of the body can be supported; then movement will stop. Let the total area of real contact be A when this happens (Fig. 9.2(d)). Then A is determined by

$$\frac{Mg}{A} = S_Y, \tag{9.1}$$

Typically, S_Y is of the order of 10^{10} dyn/cm^2 for a metal. If the mass of the body is 100 g, Mg is 10^5 dyn and A is of the order of 10^{-5} cm^2—though a body weighing 100 g is likely to have dimensions of a few centimetres.

Now imagine a horizontal force F to act on the upper body. The contacts experience a shearing stress, Fig. 9.2(e), equal to this force divided by the area tangential to it, that is F/A. They can withstand this deformation until the stress reaches a limiting value called the *shear strength*, denoted by S_S, when the contact snaps. In a real situation there are of course many contacts which break at different times and reform elsewhere, but the force required to move the body is given by

$$\frac{F}{A} = S_S. \tag{9.2}$$

Eliminating A from these two equations,

$$\frac{F}{Mg} = \frac{S_S}{S_Y} = \mu \tag{9.3}$$

where μ is called the coefficient of static friction, the ratio of the horizontal force acting on the body to its weight. Thus we have shown that μ is independent of the size or shape of the bodies and is equal to the ratio of the shear strength to the yield strength. Roughly, this is equal to the ratio of the rigidity modulus to Young's modulus, and this from the table in section 3.5 is roughly $\frac{1}{2}$. Indeed it is observed that the coefficient of friction between metals is of this order.

In addition to the welding and snapping of the regions of contact, the coefficient of friction is affected in practice by a number of other effects which are of technical importance. For example, the ploughing of one asperity through another as the body moves sideways can be important. Lubricants and films of oxide on the surface can have a profound effect on friction. Further, when two different metals are in contact, the fact that one metal melts more easily than the other can alter the details of the mechanism. We will content ourselves however with having related the phenomena at the surface with the properties of the bulk material.

9.2   BRITTLE MATERIALS

Glasses and ceramics (like pottery and bricks, which are mostly metallic oxides) are brittle materials. They have disordered, non-crystalline structures: Fig. 9.3 is an attempt to picture some neighbouring atoms in such a solid. The interatomic bonds are highly directional in nature, being of the electron-sharing type (section 2.1.4). The coordination number of each atom is small and the structure more or less rigid. As mentioned previously (section 2.2.4), amorphous solids can be pictured as liquids 'frozen' into a particular disordered configuration which does not change with time. For a molecular solid, like solid argon in a non-crystalline state, the molecules are arranged as in Fig. 2.4; the coordination number is high, the bonds directed at all angles. Amorphous regions in metals have similar structures. Fig. 9.3 and Fig. 2.4 resemble one another

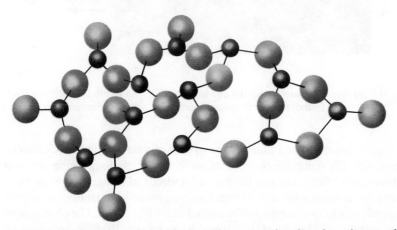

Fig. 9.3. Atomic structure of an oxide glass. The atoms or ions have been drawn rather small for clarity. The lines joining them represent directional covalent bonds.

in that both are disordered, but they differ in that one lattice is much more rigid than the other.

To explain their stress/strain curves, it has been proposed that brittle materials contain cracks, which may be of microscopic or even atomic dimensions, and which may either be wholly inside the material or else originate on the surface. Such imperfections weaken the material and it breaks when one crack suddenly spreads. A familiar example of such behaviour is the way one can break glass in a controlled way by first scoring the surface with a diamond and then bending the glass so as to open the crack.

We can simulate the stresses round a crack of known shape but of large size, and study them using an optical method. Materials like glass or plastics when deformed exhibit double refraction. This means that the refractive index for visible light is different for light polarized parallel to the direction of the stretching, from that for light polarized at right-angles to it. Using polarizing plates and a quarter-wave plate in a way which is described in standard texts on optics, the strain in a photoelastic material can actually be seen and measured through the intensity of light it transmits. In particular, lines of constant strain can be made visible as dark lines, interference fringes, on an illuminated background.

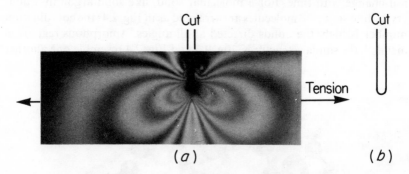

Fig. 9.4. (a) Stress pattern round the end of a cut in a slab of photoelastic material. (b) End of the cut showing semicircular end of radius ρ.

Figure 9.4(a) shows a photograph of part of a slab of plastic (Columbia resin, CR 39) 6 mm thick, 19 mm wide and 7 cm long being stressed horizontally with a tension of several hundreds of kilograms. A parallel-sided cut, 0.4 mm wide and 11 mm long was made in it, transversely, with the ends carefully machined to be semicircular Fig. 9.4(b). The photograph concentrates on the end of the cut and it can be seen that the dark lines all crowd in there.

In some preliminary experiments with a specimen of plastic of the same thickness but with no cut in it, the optical components were set so that the specimen was uniformly dark when there was no tension on it. When the tension was increased, the field lightened gradually and then went dark again; this occurred with a tension of 1,200 kg/cm². Subsequently, the field went dark again every time the tension was increased by about this amount. Now each fringe on the pattern represents a line of constant strain: we can therefore say that if Hooke's law holds the tension at all points along one fringe differs from that at points in the next fringe by 1,200 kg/cm². The fact that all the dark lines crowd together therefore means that the tension varies very rapidly from point to point, and that it is a maximum at the end of the cut.

From the intensity at the top and bottom of the specimen, the applied tension was estimated to be about 600 kg/cm². There are 7 fringes between these regions and the end of the cut, so that at that point it is 8,400 kg/cm² higher. Hence the tension at the end of the cut is about 15 times as great as the tension far from it.

The theory of these stress/strain patterns in irregularly shaped bodies is complex: we will however quote the result that if a transverse cut is of total length l and the radius of the end is ρ, the stress in the neighbourhood of the end of the cut is enhanced by a factor

$$2\sqrt{\frac{l}{\rho}}.$$

Here, this is $2 \times (11/0.2)^{1/2} = 15$, in good agreement with the measurements on the photoelastic specimen.

It is obvious, then, that a crack inside a solid with a small radius ρ at the end—that is, a sharp pointed crack—*concentrates the stress* at the end. Whereas the tension far from the crack may be quite small, well within the tensile strength of the material, the stress at the point may become too great and the material will fail there. This occurs when the stress s is given by

$$s_m = s \cdot 2\sqrt{\frac{l}{\rho}} \qquad (9.4)$$

where s_m is the expected breaking stress, in the absence of cracks. As the crack extends, l becomes larger and ρ remains the same or gets smaller, so that it continues to grow.

There are two points to note. The first is that cuts or cracks which are parallel to the line of the tension have no effect as stress concentrators. This is the reason for the great strength of freshly drawn fibres of quartz or glass—they can withstand elongations of several percent and their

Young's moduli are higher than that of steel so that their breaking stresses are higher than that of steel. They owe this property to the fact that when they are prepared by pulling a filament of melted but highly viscous liquid just above the melting point, all cracks become pulled out also, but parallel to the length of the fibre. Until the surface becomes pitted by chemical action through exposure to the atmosphere, they retain their strength.

The second point is that only the ratio of the length of the crack to its width enters into the stress ratio. A crack 4 Å wide and 110 Å long would be just as effective as the one in the specimen of Fig. 9.4. Now most ceramics are found to be inhomogeneous in their crystal structure when examined by X-ray diffraction. It is quite possible therefore that in a structure like Fig. 9.3 with rigid, directed bonds there may be misfitting atomic planes with few bonds crossing them. These behave like long narrow cracks, and can therefore act as stress concentrators. Glass on the other hand usually seems to break because of defects on the surface—scratches or particles of dust picked up during manufacture. The evidence for this comes from the study of the fractured surfaces. There is always a small area with a very smooth surface—so smooth that no roughness can be seen even under a microscope. But surrounding this small area, the surface is covered with small ridges. These different areas are thought to be caused firstly by the rapid tearing of the glass and secondly a forking of the crack as it accelerates. When a rod is deliberately notched and then bent and broken, so that there is no doubt where the break originated, the mirror-like surface is always found near the origin and the ridges radiate roughly from the same point. Thus we have a good method for locating the origin of any crack. All cracks seem to start from the surface, not from inside.

The great strength of brittle materials when they are compressed is explained by the closing of the cracks and notches. Glass can be toughened using this effect, by arranging that the surface layers of a sheet of molten glass cool more quickly than the inside; the inside of the sheet is the last to solidify. This has the effect of compressing the outer layers, that is, the atoms in the outer layers are closer together than those in the inner layers. Therefore when the surfaces are stressed, these tensions must first overcome the locked-in compressive stresses before any cracks can begin to open so the glass is stronger than when prepared normally. Concrete can similarly be prestressed by allowing it to harden under great pressure.

★ **9.2.1 Dynamic behaviour of brittle materials**

It is good common sense to argue that if a number of cracks begin to spread through a solid, only one of them can 'win', and that once the solid has broken the stress is immediately relieved and the other cracks must stop growing. In other words, a solid ought to break only at one place at once. It would be a gross and improbable coincidence if there were two cracks of *exactly* the same rate of growth so that a solid broke at two places at once.

Yet if one bends a glass rod (say 25 cm long and 3 mm diameter) rapidly but firmly so that it breaks, it may fly into several pieces, three or four or more (Fig. 9.5.). This is seemingly impossible behaviour. But it is not difficult to demonstrate that the cracks occur one after the other and not simultaneously, so that the commonsense argument is not incorrect. Similar things happen when a rod is broken by straight tension.

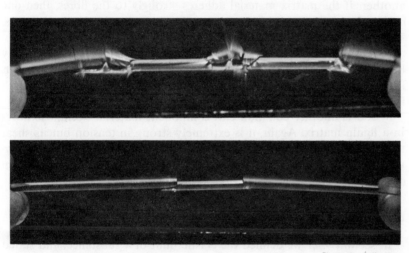

Photos by Colin King

Fig. 9.5. Flash photographs of two different glass rods breaking under the action of shear—the rods were bowed upwards.

The key to these phenomena is the presence of stress waves which are propagated along the rod. In a rod broken by bending, they start as a flexural pulse which whips back and forth along the rod. This is a complicated motion in which short wavelengths travel fastest and an initially compact pulse straggles; later, the shortest wavelengths can start bouncing around from side to side inside the rod. Or if a rod is broken by pulling, then after the break the surfaces spring back and a *compression* pulse is propagated along each half. When it meets the far end it is reflected. If that end is free to move, being unclamped, then the pulse is reflected as a *tension* pulse. For a short time, while the incident compression pulse is passing over the reflected tension pulse, the stresses cancel, but later the tension appears—and the rod may break again, a little away from the end.

★ 9.2.2 Two-component materials

Fibre-glass is a typical two-component material. It consists of glass fibres (which in the absence of surface flaws have great tensile strength) embedded in a matrix of soft, non-brittle plastic. Consider the fibres all to be parallel and the tension to be applied in their direction. Then practically all the stress is borne by the fibres. Some may crack across, but because they are separated from each other a crack cannot spread from one to another. If the matrix material adheres strongly to the fibres, then one broken fibre will be held in place through the strength of its neighbours. The tensile strength of the material is therefore extremely high.

Its shear strength is small, however, because the fibres can be easily bent, so that aligned fibre glass is not a good structural material. However, if the fibres are tangled up and are aligned in all directions, the strength for all kinds of stress can be quite high.

Wood is another common two-component material—cellulose fibres in a lignin matrix. Again, it is extremely strong in tension but its shear strength is low. This can be improved by laminating it to form ply-wood, in which successive layers have their fibres oriented in different directions.

9.3 DEFORMATION OF DUCTILE METALS

In section 9.1.1 we defined the elastic limit of a substance, and stated that soft metals prepared as single crystals can be strained beyond their elastic limits by the action of quite small forces. Cadmium notably can be pulled to 10 or 20 times its original length.

When the surface of a single-crystal rod is examined after it has been pulled beyond its elastic limit, it is seen that its surface is covered with fine lines. These show that the material has divided itself into bands which have slipped or glided with respect to one another, Fig. 9.6. The direction of easy glide are close-packed crystallographic planes. Each section has slithered sideways, pulled over by the component of the tension parallel to the glide direction, but as a result the specimen as a whole has become longer. The elongation has taken place because each section undergoes shear with respect to the neighbouring ones.

Each unslipped section has an almost perfect crystal structure: the deformation is located inside regions which are only a few atomic planes thick. The width of the unslipped regions may be anything upwards from a few thousand atomic spacings and the steps on the surface may be a few hundred or thousand atomic spacings high, so that the structure is usually visible under a microscope.

Thus the process of deformation in a ductile metal is quite different from that which we considered in section 9.1 where we worked out the tensile strength of a material—on the assumption that the deformation was elastic and would return to zero when the stresses were removed. There, we assumed that the strain was homogeneous, that the distance between every atom and its nearest neighbours increased by the same amount. But it appears that, beyond the elastic limit, the strain is far from homogeneous: large movements take place in relatively few planes.

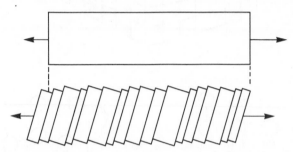

Fig. 9.6. Stretching of a single-crystal rod under tension by slipping in sections along direction of easy glide. The sheared rod is slightly longer than it was originally.

9.3.1 Dislocation lines

It is useful to study the strains in a metal where the slip has travelled only *part way* across the thickness. Fig. 9.7(a) shows this diagrammatically. The whole block, originally rectangular, is now distorted and out of true. Part of the left-hand front face has been pushed in a short distance, and an incipient step is visible on this face. With this deformation, part of the whole block has also slipped and the distortion of the atomic planes is visible on the right-hand face. The boundary between the slipped and the unslipped region is marked by a curve inside the block, shown by the shading on the slip plane. This boundary is called a dislocation line.

The strain, the displacement of an atom relative to its neighbours, is greater near the dislocation line. But elsewhere, the crystal is almost perfect.

It is these relative displacements in the vicinity of the dislocation line which we will have to study in some detail. Now when the dislocation line is curved, as in the diagram, the strains are difficult to describe. But two basic types of *straight* dislocation lines can be distinguished, and any curved line can be regarded as a combination of these.

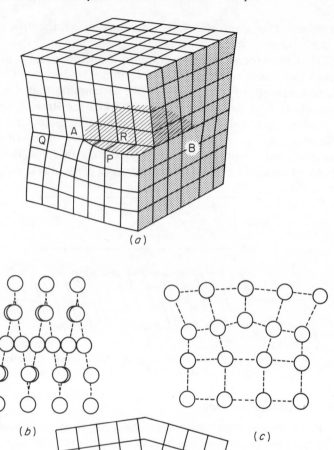

Fig. 9.7. (a) A strained cube with a slipped region bounded by the shaded arc. At A a screw dislocation emerges, at B an edge dislocation. The deformation is in the direction RP. (b) Looking down on a screw dislocation; A, P, Q, R correspond to A, P, Q, R in the cube above. In a perfect lattice the atoms would lie on separate lines, but here they lie on a helical surface. (c) Looking along an edge dislocation. (d) Small-angle grain boundary; D and D' are edge dislocations.

In the diagram, the slip is in the direction of RP. When the dislocation is a straight line parallel to the slip, as it is near A, it is called a screw dislocation. When it is perpendicular, as it is near B, it is called an edge dislocation.

The arrangement of atoms near to a screw dislocation line is shown in Fig. 9.7(b). What were originally parallel planes of atoms have now become distorted into a helical surface, like that generated by a propellor rotating slowly as it moves forward. Starting on one plane at the top of the diagram and moving from atom to atom to the bottom and then up again, one arrives without discontinuity at the next plane; in fact, there is strictly speaking only one single sheet of atoms, instead of parallel planes. The left-hand face, with the step, in Fig. 9.7(a) is the outermost layer of this sheet.

The atoms near an edge dislocation are shown in Fig. 9.7(c). In the upper half, there are five columns of atoms, in the lower half only four. Similar patterns are visible on the right-hand face of Fig. 9.7(a), and also in Fig. 2.5.

Dislocations may be produced by any irregularity during the growth of a crystal. They occur at grain boundaries, the junctions between crystals which do not fit perfectly because there is an angle between the sets of crystallographic planes. The resulting misfit can be described as a row of edge dislocations, regularly spaced, as in Fig. 9.7(d). Dislocations may also be clustered round any misfitting inclusions or foreign atoms inside a lattice.

Dislocations can be seen most easily in specimens made in the form of very thin films, say 1,000 Å thick, examined in an electron microscope capable of magnifying by a factor of the order of 10^5. Dislocation lines which run more or less vertically through the thickness are visible. Points like A and B in Fig. 9.7(a) where dislocations emerge at surfaces of crystals are chemically active, because of the high energy of atoms with the wrong coordination. Thus they are vulnerable to chemical attack by appropriate reagents. Little holes called etch pits appear at odd places on the surface. Another chemical method for seeing dislocations is the 'decoration' of the lines inside transparent crystals like silver bromide and silver chloride. These substances are used in photographic plates because, under the action of light, free silver is formed which is opaque. Not surprisingly, it is formed preferentially on the dislocation lines so that these can be seen under a microscope.

In a crystal of metal prepared under ordinary conditions, there are typically about 10^7 dislocations in any square centimetre of cross section. This sounds an enormous number, but expressed rather differently the same datum shows that such a crystal is highly regular. For if the interatomic spacing is 2 Å, there are 5×10^7 atoms in 1 cm of length; 10^7

dislocations per cm^2 means one dislocation every 3×10^{-4} cm, that is one every 15,000 atoms. For comparison, it may be noted that in the most perfect of single crystals which can be prepared the density is around 10^3 dislocations/cm^2. At the other end of the scale in heavily damaged metals where the crystal structure has been broken up there might be 10^{12} dislocations/cm^2, one every 50 atoms. Even in the most disordered crystals, the assumption that most of the atoms are in their regular lattice sites is a valid one for many purposes.

9.3.2   Movement of dislocations

Having described what dislocations are, we can now show how they can produce slip in a crystal under the influence of small forces—tensions which are small compared with the elastic moduli.

The essential process is the *movement* of a dislocation through the lattice. We will concentrate on edge dislocations. Similar arguments apply to screw dislocations but it will be left as an exercise for the reader to construct these.

Imagine a block sheared as in Fig. 9.8. A dislocation is created on the left, runs from left to right and emerges from the block. The net result is that the top half of the crystal has sheared over. By itself, this process causes only a very small slip, only one atomic spacing high. But it is the basic process which is responsible for the inelastic extension of metals under small forces. Given 10^7 dislocations per cm^2, a movement of each one of them would add a 2 Å step to one side, and if all of them moved a 1 cm cube would shear over a distance of 2 mm, a 20 % strain.

We can now look at Fig. 9.6 with much more understanding than before. One would guess at first sight that each slip band has slipped bodily across, like a penny in a pile which has been pushed askew. But now we can see that each section has *not* been moved like a rigid body—dislocations have moved across instead—and when we say that a dislocation has moved, Fig. 9.9 shows that all we mean is that all the atoms in the slip plane have moved a small fraction of an atomic spacing, one after the other.

It is not difficult to see qualitatively that dislocations, once created, can move through a lattice very easily. Not only are the individual displacements of the atoms very small, which implies small activation energies, but when one atom gains potential energy (by moving a little further from its neighbours) another loses an almost equal amount (by moving closer). Therefore during the movement the potential energy of the system hardly changes.

Quantitatively, however, the mobility depends critically on the geometry of the lattice. The dislocations which we have pictured have all been *narrow* ones, in the sense that there are only two or three atoms in a line which are badly displaced from their regular positions. The number ω of such atoms in a line is a useful measure of the width. But it is possible to produce *wide* dislocations, as in Fig. 9.9(b). Here there are many more displaced atoms in a line but the change of position of each one when the dislocation moves is much smaller than before so that wide dislocations are more mobile than narrow ones. A detailed analysis shows that if s_d is the stress required to move a dislocation and K is one of the elastic moduli, then in order of magnitude

$$\frac{s_d}{K} \sim e^{-2\pi\omega}. \tag{9.5}$$

$e^{2\pi}$ is equal to 500 so that the width ω of the dislocation should have a dominant effect on the yield strength of a material.

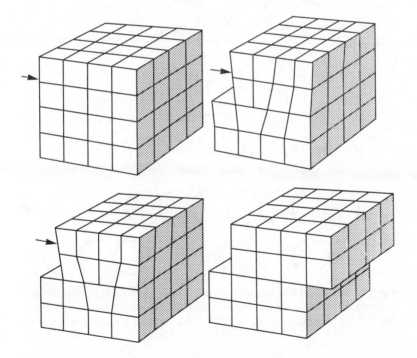

Fig. 9.8. Shearing of a block by the movement of a dislocation across it.

Wide dislocations are favoured by large spacings between atomic planes; such planes themselves are close-packed crystallographic planes. Hence one would expect that dislocations can move most easily in close packed crystallographic planes. This result was observed experimentally in metals and has already been mentioned in section 9.3.

Wide dislocations are also most easily formed in substances like metals where the bonds are non-directional. Indeed, very pure metals are extremely soft. However, another effect enters as soon as a metal has been repeatedly deformed: it is observed that its hardness increases. This is called work hardening. It is interpreted to mean that the metal has become full of dislocation lines—and it is difficult for one line to pass through another. The lines repel one another because of the stresses round them; thus the movement of dislocations is impeded, and the metal hardens. Ordinary metals also owe some of their hardness to another effect, the presence of impurities which can also impede the motion of dislocations.

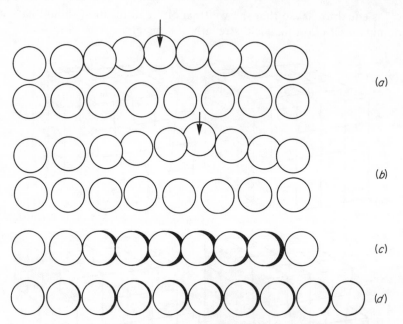

Fig. 9.9. (a) and (b): Dislocation moves one atomic space. (c) Relative positions of atoms in top line in (a) and (b), showing that each atom moves only a small distance. (d) Same diagram for a wider dislocation: ω is about 8 whereas ω is about 5 or 6 in (c).

Hard metals like steel owe their combination of ductility and hardness to the judicious introduction of other elements such as carbon, and of several types of crystal modifications dispersed throughout the material.

Narrow dislocations are characteristic of crystalline substances with covalent bonds, which are rigid and difficult to distort. Covalent materials in any case have high elastic moduli, but in addition we can now see why they are not ductile. Diamond, for example, is the hardest substance known.

9.4   GROWTH OF CRYSTALS

We will now discuss the rate of growth of crystals and to simplify matters we will consider growth from the vapour phase. The classic studies were done on iodine crystals; iodine has a vapour pressure of 0.1 mm at room temperature. The rates of growth were typically about 0.1 mm per hour when the vapour pressure was 1.05 times the saturated vapour pressure—that is, when the vapour was 5% supersaturated. In this section we will outline a simple theory and demonstrate that it is wrong by many orders of magnitude. We will *assume* that the initiation of a new layer of molecules on a plane crystal face is much like the nucleation of a droplet of liquid in a supersaturated vapour (see section 7.8.3). We will also assume that the crystal face is perfectly plane to begin with. If a single molecule arrives from the vapour and sticks to the plane, the number of bonds which it forms is only a small fraction of its full coordination number, about $\frac{1}{4}$. So this lone molecule probably evaporates again after a short but finite time. If, however, there is an 'island' already formed on the face (a two-dimensional droplet, as it were), the newly arrived molecule might lodge at the edge of it and then its coordination number approaches $n/2$ and there is a much greater chance of it sticking permanently; the island acts as a nucleus, Fig. 9.10. Let us for simplicity assume that the island is circular. Then we can treat it like a *cylindrical* droplet of finite radius and

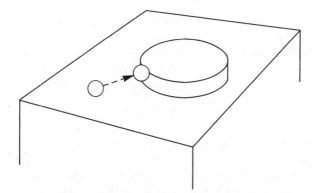

Fig. 9.10. The formation of an 'island' from the aggregation of newly-arrived molecules landing anywhere on the crystal face.

only one molecule high. Using exactly the same methods as in section 7.8.3 for spherical droplets we can calculate the critical radius for a given degree of supersaturation of the vapour, the radius of island which can grow instead of evaporating. We do this by balancing the rate of evaporation from the curved surface against the rate of arrival of new molecules which might have landed *anywhere on the crystal face and diffused* to the island. (This last is different from the corresponding conditions during the formation of a liquid droplet). Finally we can calculate the activation energy A_0 for forming such an island; the probability of a critical island forming contains the Boltzmann factor $\exp(-A_0/kT)$. The steps are as follows. Inside a cylindrical droplet of any length but of radius r, the excess pressure is γ/r. Its vapour pressure is therefore increased over that of a plane surface by the factor

$$\frac{P}{P_0} = e^{\gamma v/rkT}$$

where v is the volume of one molecule, γ the surface tension, k Boltzmann's constant; compare Eq. (7.34). The critical radius is

$$r_c = \frac{\gamma v}{kT \ln (P/P_0)};$$

compare Eq. (7.35). The activation energy A_0 for forming such an island is (following precise thermodynamic arguments) equal to one-half of the surface energy of the edge:

$$A_0 = \pi r_c d\gamma$$

(compare and contrast Eq. (7.36)) where d is the diameter of a molecule (the height of the cylinder).

Let us simplify these expressions by writing $\gamma = \frac{1}{4}\mathcal{N}n\varepsilon$ (see section 3.4) where $\mathcal{N} = 1/a_0^2$. The volume v of one molecule is a_0^3. Then:

$$A_0 = \frac{\pi}{16}\frac{(n\varepsilon)^2}{kT \ln (P/P_0)} \tag{9.6}$$

for an island of critical radius to be nucleated in a supersaturated vapour of pressure P.

Let us now put in numbers. For a molecular crystal, let us take $n = 10$, $\varepsilon = 0.1$ eV and for room temperature $kT = \frac{1}{40}$ eV.; whence

$$A_0 \approx \frac{10}{\ln(P/P_0)}\text{eV}.$$

Suppose the area of face is 1 mm square, 10^{-2} cm^2. At a vapour pressure of 0.1 mm, the number of molecules arriving at unit area, $\frac{1}{4}n\bar{c}$, is 10^{18} per second. Further, the critical radius is always small and a new nucleus might be formed anywhere on the crystal face. Thus the number of critical nuclei formed per second is equal to the number of molecules arriving per second on the whole face multiplied by the probability that one of them has enough energy to form the nucleus:

$$10^{16}\exp(-A_0/kT) = 10^{16}\exp[-400/\ln(P/P_0)].$$

Consider a crystal growing at 0.1 mm per hour, which (for $a_0 = 3$ Å) is 3×10^6 layers per hour or 1,000 layers per second. For this

$$10^3 = 10^{16}\exp[-400/\ln(P/P_0)].$$

This tells us the supersaturation required; it gives $\ln(P/P_0) = 13$, which means that the vapour pressure should be e^{13} or 5×10^5 times supersaturated. Experiments gave 5% supersaturation for this rate of growth. Thus the theory is badly wrong.

The weak point of this theory is that after an island has grown and covered the whole crystal face, it produces a perfect plane; a finite time, very long on the molecular time-scale, has to elapse before another fluctuation occurs which is big enough to form another island. In order to explain the observed rate of growth, we need to postulate some sort of step against which newly arrived molecules can stick themselves, but a step which remains in existence even after the face has been covered with a new layer.

9.4.1 Growth spirals

The point of emergence of a screw dislocation at a crystal face provides just such a step—Fig. 9.11. We can imagine atoms arriving more or less uniformly all along the line and the crystal can grow. But the step does not advance as a straight line. Near the centre of the screw there are some sites where the step is not of the full depth, and the atoms lodging there soon convert the end of the straight ledge into a tight spiral. This spiral then grows upwards like a ziggurat, and the crystal grows as the turns of the spiral sweep round over the crystal surface—and they continue to advance however much the crystal grows. This is ultimately a consequence of the fact that there is only one sheet of atoms in a crystal containing a single screw dislocation.

Spiral markings can be found on many crystals under a microscope, Fig. 9.11(e). The faces of an average crystal have many screw dislocations emerging on them, and no point of a face is more than a few hundred atomic distances from a spiral step. This fact obviously increases the chance of permanently capturing a molecule from the vapour. But at first sight, it might appear that such a molecule would have to land right on the step, or within one atomic distance or thereabouts from the step, in order to be captured. But this is not so: we can show that a lone molecule, landing far from a step, sticks for a comparatively long time and diffuses an appreciable distance over the surface before being thrown off again. In its wandering it can meet a step and stick there, so that the probability of capture is greatly enhanced.

We show this as follows. For an isolated molecule on a flat plane, the number of bonds it forms is about $\frac{1}{4}n$ where n is its full coordination number. The probability of its evaporating therefore contains the factor $\exp(-\frac{1}{4}n\varepsilon/kT)$.

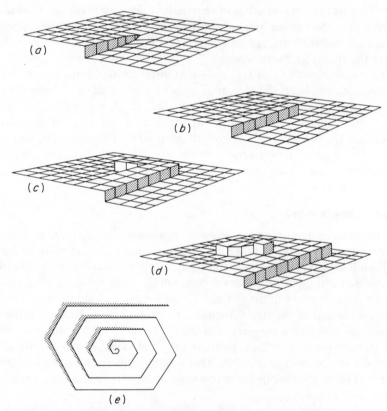

Fig. 9.11. First four pictures show the early stages in the growth of a spiral step on a crystal face round the point of emergence of a screw dislocation. (e) shows the appearance of a growth spiral in the electron microscope: the steps are typically 100 Å to 1,000 Å across.

Now the molecules on which it is sitting are vibrating at the Einstein frequency (see section 3.6.1), which we call v_E. One would guess that there is a maximum probability of the lone molecule being thrown off once per vibration. Therefore the chance per second that the molecule evaporates is $v_E \exp(-\frac{1}{4}n\varepsilon/kT)$. With reasonable numbers for a molecular solid, the 'sticking time' is about 10^{-8} s. The molecule can jump to a neighbouring site without evaporating, and the energy needed to do this is very small indeed compared with ε. One can imagine the molecule sitting in a dimple on the surface and the activation energy required to make it jump to a neighbouring dimple is obviously very small. Since the corresponding Boltzmann factor is almost equal to unity, it probably jumps once every vibration, so that its jump frequency is v_E. Therefore it performs $\exp(\frac{1}{4}n\varepsilon/kT)$ jumps before evaporating. Each jump moves a distance a_0, but it performs a random walk. The net distance it goes in s steps is therefore only $a_0\sqrt{s}$ (see section 6.3). Hence it diffuses an average distance

$$a_0 \exp(\tfrac{1}{8}n\varepsilon/kT)$$

(where the effect of taking the square root of the exponential is to alter the $\frac{1}{4}$ to $\frac{1}{8}$). For $n\varepsilon = 1$ eV, $kT = \frac{1}{40}$ eV, this is about 150 atomic diameters.

This is comparable with the distance between the spiral steps which are to be found on the average crystal face. Therefore we can say that practically every molecule that arrives from the vapour is eventually captured. Of course the crystal also loses molecules by evaporation. The number lost per cm^2 in one second is $(\frac{1}{4}n\bar{c})$ evaluated at P_0, the equilibrium vapour pressure (because this must be equal to the rate of arrival of molecules, in equilibrium). If, however, the crystal is exposed to vapour at supersaturated pressure P, the rate of arrival of molecules is $(\frac{1}{4}n\bar{c})$ evaluated at P. Therefore the net rate of gain of molecules is

$$\frac{1}{4}n\bar{c}\left[\frac{P}{P_0} - 1\right]$$

per cm^2 in one second, where the first factor is about 10^{18} for iodine vapour at room temperature. Putting in numbers, the supersaturation required for a 1 mm square crystal face to advance at a rate of 0.1 mm per hour is 10%—in good agreement with the 5% observed.

This detailed study of the mechanism of crystal growth is not incompatible with the account given in section 8.1.3. If screw dislocations inside a growing crystal are oriented in all directions at random, then the growth will proceed as in Fig. 8.6.

★ **9.4.2 Single-crystal whiskers**

Under the right conditions, crystals can grow in the form of tiny whiskers, typically a few thousand atoms in diameter and anything up to millimetres in length. NaCl and KCl can, for example, be grown out of the sides of big crystals by exposing them to the vapour at temperatures of the order of 1,000° K. X-ray examination shows that these filaments are single crystals. It is thought that they are produced when the outer turns of the growth spirals on the surface of the big crystal are stopped from advancing because foreign impurity atoms have lodged there. This allows only the tight central turns of the spiral to grow upwards and the result is a filament. Single-crystal whiskers are extraordinarily strong. For example, the specimen of iron capable of withstanding a 4% elongation which was elastic and reversible (see section 3.7.1) was a single-crystal whisker. This strength is in marked contrast to the ductility of *large* single crystals of metals. But we expect a whisker to have only one single screw dislocation running down the middle of it. Such a line defect does not weaken a lattice when tension is applied parallel to it—in much the same way as a crack does not weaken a specimen of brittle material if it is parallel to the stress (section 9.2). The observed strength of single-crystal whiskers therefore confirms the proposed explanation for their growth. It should be mentioned for completeness that other mechanisms of growth are possible—some whiskers actually seem to grow from the base, pushing their way upwards.

★ **9.5 POINT DEFECTS**

We have described dislocations in solids, but there exists a much simpler kind of imperfection, the point defect. This is the single atom (or molecule or ion) which has been displaced from its lattice position. If it has lodged in a nearby site, wedged in between a near set of neighbours which have to move a little out of the way in order to accommodate it, it becomes an *interstitial* atom; it leaves a *vacancy* behind, Fig. 9.12. Vacancies can exist by themselves without any corresponding interstitial atom: in that case we can regard the displaced atom as having been removed to the surface of the crystal.

Vacancies are important because their presence greatly enhances the speed of diffusion of atoms in solids, and this is because the activation energy for one to jump one atomic space is not large. We can follow a line of argument suggested in section 7.2 and regard liquids as resembling highly disordered solids containing many vacancies and interstitial atoms; we can then gain new insights into diffusion mechanisms in liquids and thence into other transport processes.

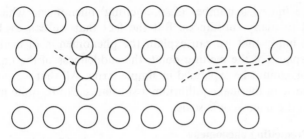

Fig. 9.12. Atoms removed to an interstitial position and
to the surface, leaving vacancies behind.

★ 9.5.1 Concentration of vacancies

First we will estimate the number of vacancies in a solid as a function
of temperature. Atoms can jump out of place as a result of thermal agitation so that the displacing of an atom is thermally activated. Let us
compare the energy of a perfect lattice with that of a lattice containing
one single vacancy, with the displaced atom removed to the outer surface.
We can imagine the atom to be cut out of the lattice by breaking n bonds
(n the coordination number). This requires energy $n\varepsilon$. The atoms around
the vacancy then relax a little towards one another and this reduces their
energy. Finally, putting the one atom back on the surface reduces the
energy a little further still. The net result is that the energy of the whole
crystal is increased by something like $\frac{1}{6}n\varepsilon$. For solid argon this is about
0.02 eV, for sodium chloride it is about 3 eV, a much larger value because
of the long range forces.

Let the increase of energy of the crystal due to the presence of one
vacancy be denoted by ΔE. The Boltzmann factors for the crystal with
one vacancy and for the perfect crystal can be written down; the ratio of
probabilities of the two states is $\exp(-\Delta E/kT)$.

Thus for a solid in equilibrium the fraction of vacancies is
$\exp(-\Delta E/kT)$. This function increases with temperature and for
a solid reaches its greatest value when the solid melts. We therefore expect some warning of the onset of melting to be given in
the last 10° or 100° before the solid melts, in the form of large numbers of
vacancies. In section 8.3 we noted that the amplitude of atomic vibrations
in a solid at the melting point was surprisingly small—one-tenth of the
interatomic spacing in molecular solids, one-thirtieth in ionic solids. We
will now be able to show that under these same conditions, vacancies are
produced in large numbers. The smallness of the amplitude at melting
therefore appears a little less mysterious, because we can accept that a
lattice which is full of holes can collapse more easily than a perfect one.

There are indeed a number of 'premelting' phenomena which can be observed in many substances. But the fact that we can explain these does *not* imply that we can explain the phenomenon of melting itself. The sharpness of the melting point, the suddenness of the onset of the transition, cannot be explained in simple terms. All that we can show is that because of thermal equilibrium a substance can be quite inhomogeneous on the microscopic scale.

★   ### 9.5.2   Premelting phenomena

For rough calculations on molecular substances and metals near their melting point let us take the melting temperature T_f to be half the critical temperature T_c :

$$kT_f = \tfrac{1}{2}kT_c = \tfrac{1}{2}\varepsilon$$

(see section 8.3). Then $\Delta E/kT_f$ must be about 4 for a close packed lattice with $\mathrm{n} = 12$. Now $\exp(-4) = \frac{1}{50}$; thus at the melting point we would expect about 2 % of the atoms to be displaced. This means that about 1 atom in 4 linearly is out of place and really it represents a highly disordered lattice. Vacancies are therefore very close to one another and the energy required to produce a new one is thereby reduced. So our estimate is, if anything, likely to be too low. Further, we have not taken the expansion of the lattice into account; this would make it easier for atoms to slip through their neighbours. In other words, the activation energy should decrease with temperature. Again, this means that we are underestimating the number of vacancies. Since we are dealing with exponentials, we *may* be wrong by orders of magnitude. In fact it will emerge that, numerically, any simple theory of this sort is not very useful.

In thermal equilibrium there must be vacancies present. Perhaps unexpectedly, a perfect crystal cannot exist at a finite temperature. Vacancies in turn affect the specific heat. In other words, if heat energy is put into a body, not all of it goes into raising the temperature but some goes into creating defects. The specific heat is thereby increased. We will calculate this with our simple theory.

If there are N atoms, $N \exp(-\Delta E/kT)$ are displaced and each one has an extra energy ΔE in addition to the $3kT$ it has (if it is a monatomic molecule) in the perfect lattice. So the mean energy is

$$\bar{E} = 3NkT + \Delta E \cdot N\, e^{-\Delta E/kT}. \tag{9.7}$$

The specific heat is given by

$$C_v = \frac{d\bar{E}}{dT} = 3Nk + Nk\left(\frac{\Delta E}{kT}\right)^2 e^{-\Delta E/kT}. \tag{9.8}$$

At the melting temperature, $(\Delta E/kT)$ is about 4, so that the extra term is about $\frac{1}{3}R$ which is appreciable. We have just seen that this is likely to be an underestimate.*

Some such rise, of about $\frac{1}{2}R$, is present in the curve for argon, Fig. 5.10(b), but it is difficult to disentangle it from the variation at low temperatures due to quantization of the lattice energy. A comparable rise, of about $\frac{1}{2}R$, is present in the specific heat of krypton (melting point 115°K) which is clearly separated from the fall at low temperatures; but the difficulty with this substance is that the expansion coefficient is not known with sufficient accuracy to allow C_v to be calculated. Exact comparisons are therefore not possible for either of these two simple substances. Potassium shows a similar rise before it melts at 335°K, Fig. 9.13.

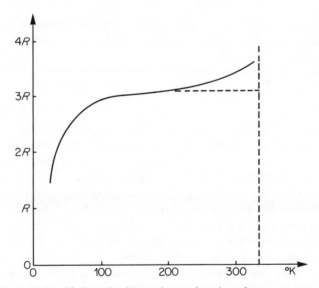

Fig. 9.13. Specific heat C_v of potassium as function of temperature.

For ionic solids, we have already mentioned (section 8.3) when discussing NaCl that small positive ions can probably slip more easily through a vibrating lattice than larger negative ions; in other words, their activation energy should be small. The concentration of vacancies is of course very sensitive to the exact value of the activation energy and as a result, extreme variations in concentration occur from substance to substance. In NaCl, it is very small. In AgBr, however, it

* The term is of the form $x^2 . e^{-x}$ and this goes through a maximum when $x \approx \frac{1}{2}$. The decrease of ΔE with temperature alters this to an increasing function.

is about 2% at the melting point, although the crystal structure is the same as for NaCl and the ratio of radii is similar. However in AgBr the bonds are not wholly ionic in character but partly covalent; perhaps this decreases the activation energy significantly. The effect of the presence of 2% of vacancies is very marked. The rise in specific heat is large, the value just below the melting point being almost three times the expected equipartition value for the perfect lattice.

Another premelting phenomenon will also be mentioned. The volume becomes greater when vacancies are formed. If there is no relaxation of the atoms round a vacancy, each one adds a volume v to the volume of the crystal. The total increase of volume due to this effect is therefore

$$\Delta V = Nv\,e^{-\Delta E/kT}.$$

Nv is equal to the molar volume of the substance. This expansion must be added to that due to the anharmonicity. The extra term in the expansion coefficient is

$$\frac{d}{dT}(Ne^{-\Delta E/kT}) = N\cdot\frac{\Delta E}{kT^2}e^{-\Delta E/kT}.$$

Let us put $\Delta E/kT$ equal to about 4 at the melting point. For solid argon, T_f is about $80°$K and since $\exp(-4)$ is roughly 0.02, the extra term in the coefficient is about 1×10^{-3}. Now it is observed experimentally that the expansion coefficient at the melting point is 1.8×10^{-3} so that our simple theory implies that over 50% of the measured effect is due to the vacancy contribution. But we have already seen (in section 8.4.2) that the Grüneisen ratio (whose constancy implies that only the anharmonicity of lattice vibrations contributes to the expansion) decreases only by about 15% as the melting point is approached in solid argon. Thus our value for the vacancy contribution must be an overestimate. In turn, this can only mean that the extra volume created by a vacancy is much less than the volume of one atom. It appears in fact that the neighbours relax a good deal and only about $\frac{1}{4}v$ is added per vacancy as the melting point is approached.

★ **9.6 DIFFUSION IN SOLIDS**

We usually consider every atom in a solid to be bound to its own lattice site and never to move from it. But this is not so; diffusion can take place in solids, albeit very slowly. An atom in a perfect lattice can change its position by jumping to an interstitial position and thence to a neighbouring interstitial position and so on. Each jump requires a high activation energy and so this process is comparatively rare. However, in an imperfect lattice where there are defects, diffusion is much easier. An atom can jump into a vacancy much more easily than into an interstitial position: the process of squeezing past its neighbours into an already existing hole requires rather less energy than that needed to create the vacancy initially —not *much* less, because of the relaxation of the atoms round the vacancy. After the vacancy has been occupied, another is left behind, Fig. 9.14, and the process can happen again. This mechanism must dominate the diffusion process and is the only one we will consider.

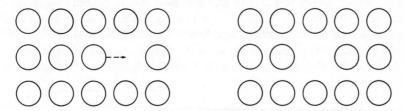

Fig. 9.14. The movement of a vacancy caused by an atom jumping into it.

The probability of an atom at A being able to jump into a neighbouring site B is equal to

(probability of a vacancy existing at B)
$\times$ (probability of jump into vacancy).

The first of these two factors we have already calculated as $\exp(-\Delta E/kT)$ where ΔE is the energy required to create the vacancy.

The second factor we can calculate by noting the resemblance to the situation shown in Fig. 5.12(*a*) and (*b*), namely that in order to jump into a hole, an atom has to squeeze past an intervening position of higher potential energy. We need to know this activation energy. Let us call it ΔE_J, which it would be reasonable to expect to be about half of ΔE. The Boltzmann factor is $\exp(-\Delta E_J/kT)$.

The lattice is vibrating at about the Einstein frequency ν_E, and one would expect an atom to have a maximum chance of jumping once per cycle. So the *number* of jumps per second is

$$\nu_E . e^{-\Delta E/kT} . e^{-\Delta E_J/kT}$$

In a time t it makes

$$\nu_E . e^{-(\Delta E + \Delta E_J)/kT} . t$$

jumps. Each one moves the atom a distance a_0. But the path is a random walk and the net distance moved in s jumps is $a_0\sqrt{s}$. Hence

$$\text{distance moved} = a_0(v_E e^{-(\Delta E + \Delta E_J)/kT} \cdot t)^{1/2}. \tag{9.9}$$

Now when we discussed the connection between the random walk problem and diffusion, in sections 6.2 and 6.3, we showed that the mean distance travelled in a medium of diffusion coefficient D in time t is, neglecting numerical constants, $(Dt)^{1/2}$. We can therefore say that the coefficient of self-diffusion in a solid is given by

$$D \approx a_0^2 v_E \, e^{-(\Delta E + \Delta E_J)/kT}. \tag{9.10}$$

Our theory therefore predicts that diffusion should become more rapid as the temperature is raised. This is in complete contrast to diffusion in gases, which becomes more rapid when the temperature is lowered. We will see that this prediction agrees with observation and that the form of the variation with temperature is correct—but quantitatively, agreement is not good.

★ **9.6.1   Comparison with experiment**

In section 6.2 we described two solutions of the diffusion equation which are the bases of all measurements of coefficients of self-diffusion. One was used in a set of experiments to measure D for copper. The idea was to follow the diffusion of the radioactive isotope ^{64}Cu through a single crystal of ^{63}Cu. The isotope was converted into copper nitrate and then electro-deposited on to a flat surface of the single crystal as a layer about 50 Å thick. ^{64}Cu, the only available radioactive isotope, has a half-life of only 13 hours. This meant that the temperature had to be raised so that it diffused a measurable distance in a time not too long compared with 13 hours. The specimen was therefore kept at as high a temperature as possible, in the region of 1,000°K, not very far below the melting point, where D is of the order of 10^{-11} cm^2 sec^{-1} so that in 1 day (10^5 s), the mean depth of diffusion $\sqrt{(Dt)}$ was about 10^{-3} cm. To find out how far the isotope had actually travelled, the specimen was put in a precision grinding machine and extremely thin layers ground off the surface. The accuracy of this operation was about 10^{-4} cm. The amount of radioactive tracer in the metal taken from each layer was determined by counting the beta-ray activity. Knowing the depth of the layer below the original surface and allowing for the decay of radioactivity with time, D could be calculated. A plot of $\log D$ against $1/T$ is a straight line, as predicted, Fig. 9.15.

Measurements were also made on solid argon by following the progress of the *stable* isotope ^{36}A through large crystals of ^{40}A. The isotope was introduced into the vapour—at the temperature chosen, the vapour pressure is large—and some atoms entered the solid and diffused inwards. Thus the concentration of ^{36}A in the vapour decreased as time went on, and this was followed by continuously analyzing the vapour with a mass spectrometer. In a typical run, the vapour pressure was several centimetres, the initial concentration of ^{36}A was 12% and it decreased to about half in an hour.

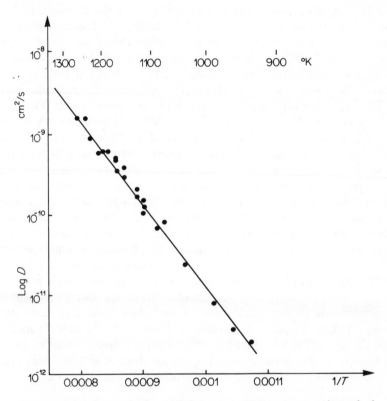

Fig. 9.15. Plot of log D against $1/T$ for copper. The temperature is marked along the top. Data from Kuper and others, *Phys. Rev.* **96**, 1224 (1954).

Diffusion can take place rapidly along grain boundaries, since these are liable to be regions with large numbers of defects. Diffusion through polycrystalline specimens is therefore likely to be spuriously large, and in all these experiments the specimens were examined to make sure that they were single crystals.

For metals, the experimental result is indeed of the form

$$D = D_0 \, e^{-A_0/kT} \tag{9.11}$$

where the constant D_0 and the activation energy A_0 are typically of the orders of magnitude 1–$10\ \text{cm}^2/\text{sec}$ and $1\ \text{eV}$ respectively. (Copper, for example, from Fig. 9.15 has $D_0 = 0.5\ \text{cm}^2/\text{s}$, $A_0 = 2.0\ \text{eV}$). These magnitudes give $D \sim e^{-40} = 10^{-17}\ \text{cm}^2/\text{s}$ at room temperature, which means diffusion through a mean distance of $1\ \text{mm}$ in 10^{15} seconds, 30 million years, at room temperature.

For comparison, our expression gives $a_0^2 \nu_E$ for the constant, and we can take a_0, the lattice spacing, to be $2.5\ \text{Å}$ and ν_E, the Einstein frequency, to be of the order of 10^{13} cycles per second. Hence we would predict $D_0 \sim 6 \times 10^{-3}\ \text{cm}^2/\text{sec}$, which is 100 times too small. For the activation energy, we would expect about 3ε or $0.4\ \text{eV}$, which is a factor 2 or 3 too small. (This has a catastrophic effect on the value of the exponential term.)

The experiments on argon reveal even worse discrepancies. They give $D_0 \sim 4\ \text{cm}^2\ \text{sec}^{-1}$, $A_0 \sim 0.11\ \text{eV}$. For an atomic spacing of $2\ \text{Å}$ and an Einstein frequency of 10^{12} cycles per second we would expect D_0 to be a factor 10^4 smaller and an activation energy 3 times smaller than that observed. It is however possible to advance plausible reasons for these discrepancies. The activation energy can be expected to decrease if the mean distance between atoms is increased; this happens as the temperature is raised. We ought therefore to get somewhat better agreement with observation if we compared our predictions with measurements at *constant density*—measurements, that is, at high pressures which counteract the thermal expansion. We return to this point in section 9.7.3. For solids, this would require the application of extremely high pressures and no measurements have been made. But in the absence of suitable measurements we can postulate some sort of decrease of the activation energy with temperature. A decrease by a factor 2 by the time the melting point is reached would indeed greatly reduce the discrepancies by orders of magnitude. However, we will not pursue this but will be content that the *form* of the expression for the diffusion coefficient is correct.

★ ## 9.6.2   Diffusion and electrical conduction in ionic crystals

A most interesting related phenomenon is that ionic crystals, usually considered to be insulators, can conduct electricity because of the diffusion of ions under the influence of an electric field. (In metals the same phenomenon occurs but it contributes a negligible current compared with that carried by the electrons: in ionic substances the movement of the ions is the only possible process.)

The potential energy of a vacancy or an interstitial ion as it moves through the lattice is shown in Fig. 9.16(a). The situation is analogous to that in Fig. 9.14 except that we expect the changes of energy to be higher than those in a lattice of neutral atoms because of the electrostatic attractions and repulsions. When an electric field acts across the crystal, the energy of the ion is changed. If a voltage V acts between the faces of a slab of thickness l, the electric field is V/l and the potential energy of a charge e coulombs changes by Vea_0/l when it moves a distance a_0. If V/l is expressed in volts/metre and a_0 is expressed in metres then the energy is measured in joules. The potential energy of an ion or a vacancy as a function of position is then shown in Fig. 9.16(b).

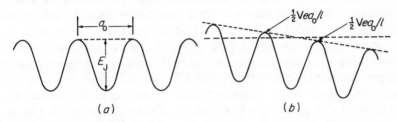

Fig. 9.16. (a) Potential energy of a vacancy or interstitial ion at different positions in a lattice. (b) The same in the presence of a potential gradient.

If the vacancy is in a potential well, the height of the barrier it must overcome in order to jump 'downhill' in the direction of the field is $(\Delta E_J - \frac{1}{2}Vea_0/l)$ where a_0 is the interatomic spacing. To jump in the opposite direction against the field, the activation energy is $(\Delta E_J + \frac{1}{2}Vea_0/l)$. Therefore the probabilities of jumping in the two directions are respectively proportional to

$$\exp -(\Delta E_J - \tfrac{1}{2}Vea_0/l)kT \quad \text{and} \quad \exp -(\Delta E_J + \tfrac{1}{2}Vea_0/l)/kT$$

which makes it more probable that the ion will move in the direction of the field than against it. In other words, an electric current i flows, proportional to the difference between the two probabilities:

$$i \propto \exp(-\Delta E_J/kT)[\exp(\tfrac{1}{2}Vea_0/lkT) - \exp(-\tfrac{1}{2}Vea_0/lkT)]. \quad (9.12)$$

Now Vea_0/l is very small compared with kT under normal conditions. For example, if V/l is large, a kilovolt per centimetre, then putting $e \sim 10^{-19}$ C, $a_0 \sim 2$ Å, $Vea_0/l \sim 10^{-24}$ J, whereas $kT \sim 10^{-21}$ J at room temperature. Since e^x is equal to $(1+x)$ when x is small, the current

$$i \propto \frac{Vea_0}{lkT} e^{-\Delta E_J/kT}. \quad (9.13)$$

To find a complete expression for the current, we should find the total number of ions (or vacancies) jumping in the direction of the field per second, and then multiply by the charge on the ion. The important results can however be seen at once: that the current is proportional to the voltage gradient (Ohm's law holds) and that the conductivity (which is the ratio of the current to the voltage) contains the same Boltzmann factor as the diffusion coefficient.

It should be remarked that both positive and negative interstitial ions and also positive and negative vacancies can contribute to the conductivity, but it is to be expected that the mechanism with the smallest activation energy is likely to dominate the conductivity.

We can therefore find the activation energy for diffusion of ions by two quite independent methods. We can use tracer methods to determine the diffusion coefficient, or we can find the electrical conductivity as a function of temperature. A plot of $\log D$ against $1/T$ and a plot of $(\log \sigma + \log T)$ against $1/T$ should be straight lines with the same gradient. In practice, the conductivity of many ionic solids changes by a factor 10^6 when the temperature changes by a factor 2; in other words, if we use logs to the base 10, $\log \sigma$ changes by 6 while $\log T$ only changes by 0.3. Thus it is accurate enough to plot $\log \sigma$ (by itself) against $1/T$ over a restricted temperature range and to take the mean gradient of that graph as the activation energy for electrical conduction.

Curves for NaCl are shown in Fig. 9.17. The resistivity is of the order of 10^3 ohm cm near the melting point, 10^6 ohm cm at 800° K. To measure high resistivities of this kind accurately, a high voltage is applied as a pulse and the current is measured as quickly as possible.

In the diagram, only high temperature observations have been plotted, above about 800° K ($1/T$ less than about 0.0013). Below that temperature the behaviour changes and both the conductivity and the diffusion coefficient become dependent on the grain size and the presence of

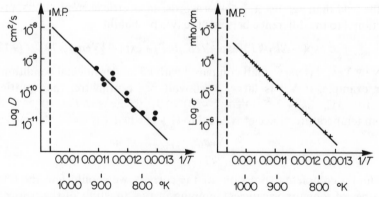

Fig. 9.17. Diffusion coefficient and electrical conductivity of NaCl as functions of temperature, ($\log_{10}$–$1/T$ plots). Data from Mapother, Crooks and Maurer, J. Chem. Phys. **18**, 1231 (1950).

impurities. Further, it is thought that positive and negative vacancies can migrate together, acting as an electrically neutral pair whose movement contributes to the diffusion process but not to the conductivity. Concentrating, then, on the high temperature region, we see that the lines are almost parallel. The activation energy is nearly 2 eV per ion, and we get a consistent picture of both the conduction and the diffusion if we assume that the small Na^+ ions are the most mobile and dominate both processes.

9.7   DIFFUSION IN LIQUIDS

We will now apply this theory, developed to account for transport through disordered solids, to liquids. This follows the plan outlined in section 7.2, Fig. 7.6.

If this approach is correct, we would expect the coefficient of self-diffusion through liquids to be of the same form as that for solids, $D = D_0 \exp(-A_0/kT)$. Of course, whenever we find it convenient we can regard liquids as derived from dense gases. This approach would lead us to expect D to be proportional to T, which is quite different behaviour. It is in fact reasonable to expect that when the density of the liquid is low, some features of its behaviour might be 'gas-like', but at high densities some features might be 'solid-like'. In the meantime, let us see whether the solid-like model of a liquid ever applies at all.

We would expect the activation energy ΔE_J for an atom to jump into a vacancy in a liquid to be much smaller than in the corresponding solid, because there is a 10% expansion in volume between liquid and solid; small changes of interatomic distance have a large effect on the activation energy. We would also expect ΔE, the energy of creation of a vacancy, to be smaller than in the solid.

The diffusion coefficient of liquid argon (radioactive ^{37}A through ^{40}A) has been measured using yet another procedure, appropriate to liquids, based on section 6.2. A narrow capillary was made, closed at both ends by needle valves. The top one was opened and pure liquid ^{40}A was condensed in; the valve was closed again. Then liquid argon containing a small percentage of ^{37}A was condensed outside. The bottom valve was opened for a known time, during which some ^{37}A diffused into the capillary. Convection, which might have disturbed the results grossly, was discouraged because the capillary was very narrow. Afterwards, the liquid in it was pumped away and the total amount of ^{37}A determined from its radioactivity. Knowing the time and the concentration in the surrounding liquid, D could be calculated. The measurements give

$$D = 6.1 \times 10^{-4} \exp(-0.027/kT)$$

where (kT) is to be measured in electron volts. The activation energy is indeed about 5 times smaller than in the solid and the form of the expression justifies our regarding a liquid as resembling a highly disordered solid in some of its properties. We will now construct a theory of the viscosity of liquids, and we will do this by showing that there is a connection between the viscosity η and the diffusion coefficient D. Since we already know D, this relation allows us to estimate η.

★ **9.7.1 Stokes' law**

As a preliminary study, interesting for its own sake, we will estimate the force on a sphere of radius r which moves at constant velocity through a liquid of viscosity η. We can concentrate on a few little specks of liquid and follow their displacements as they go past the sphere. These define *stream lines* and they represent the way the liquid flows past the sphere, or the way the sphere pushes the liquid apart in order to pass through it. Fig. 9.18 shows some of these. It implies that the flow pattern is symmetrical and that at some distance away from the sphere the liquid is hardly disturbed. This distance is called the penetration depth. We will make the assumption that the penetration depth is the same order as the radius of the sphere, which is reasonable and is what the diagram implies. We saw in section 6.1.5 that the pressure on a moving object is given by (Eq. (6.7))

$$\text{pressure} = \eta \times \text{velocity gradient.}$$

Here, the velocity gradient is of the order of magnitude v/r.

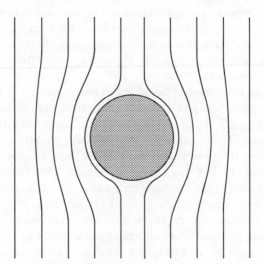

Fig. 9.18. Stream lines round a sphere moving
through a liquid.

Also implied in Fig. 9.18 is that the flow is *non-turbulent*. In the language of section 6.1.5, the kinetic energy of the mass in motion—the inertial energy—must be small compared with the energy dissipated against viscosity. This means we have to compare terms of the type $\frac{1}{2}\rho v^2$ (kinetic energy per unit volume, ρ being the density and v the velocity of the fluid in motion) with the viscous pressure $\eta v/r$ (which is also an energy per unit volume). Thus ρv^2 must be less than $\eta v/r$ to make sure that the flow is not turbulent; in other words, v must be less than $\eta r/\rho$. For a sphere of radius 10 cm moving through water for which $\eta = 10^{-2}$ gm/cm s, v should not exceed 0.1 cm/s; a spherical particle of radius 10^{-3} cm can travel at 10 cm/s and, extrapolating, a sphere of atomic dimensions can travel at 10^6 cm/s and the flow will still be non-turbulent.* With these assumptions, the pressure on the sphere is $\eta v/r$. The surface area of the sphere is $4\pi r^2$. Therefore the force on the sphere should be roughly $4\pi\eta rv$.

An exact analysis with proper regard to the flow pattern shows that

$$\text{force} = 6\pi\eta rv \tag{9.14}$$

This is called Stokes' law.

★ ### 9.7.2   Einstein's relation between viscosity and diffusion

We establish the relation between viscosity and diffusion by studying the Brownian motion of an assembly of particles. For clarity, we will consider the special situation of the fluid in the field of the earth's gravity, although it is not in fact necessary to restrict the calculation in this way. Further, for the sake of illustration we will first consider a *gas* and having shown that the results agree with previous calculations we will extend the method to liquids. We begin, then, by considering a gas in a gravitational field, as we did in section 4.4. There, we established the fact that the concentration n at a height z is given by

$$n = n_0\, e^{-mgz/kT} \tag{4.13}$$

where n_0 is the concentration at zero height and m is the mass of one molecule. The new idea is that we will regard the equilibrium as being produced by two opposing motions. First, we have the tendency of each atom to fall downwards. Secondly, opposing this, we have the tendency of each atom to diffuse along the concentration gradient (which means a tendency to move upwards, because n decreases with height). We express this balance as follows. Let us take z to be positive upwards, and consistent with this let us take the drift velocity v_z of the atom to be positive

---

* The ratio $vr\rho/\eta$ where v is the limiting velocity for the onset of turbulence is called the Reynolds number. It is about 1 for a sphere. It is much smaller for a cylinder because the penetration depth is much greater.

upwards; we will also take all fluxes of atoms (numbers crossing 1 cm^2 in 1 second) as positive upwards. Then the net flux is given by

$$J = -nv_z - D\frac{dn}{dz}.\qquad(9.15)$$

The first term on the right hand side is the number of atoms per unit area falling downwards under gravity—they are accelerating of course, but are brought to rest at intervals, and v_z is their mean velocity. The second term is the rate of diffusion down the concentration gradient.

For equilibrium, $J = 0$.

Let us next evaluate v_z, the mean drift velocity. If λ is the mean free path between collisions, the mean free *time* between collisions is $(\lambda/c)_{av}$ which is nearly equal to $\lambda/\bar{c}$, where $\bar{c}$ is the mean speed. In this time, every atom acquires an extra downward velocity equal to (acceleration) × (time), that is $g\lambda/\bar{c}$. Let us assume that after every collision, each atom is stationary. This is rarely likely to be true, because of the persistence of velocities (section 6.4.3) but it only introduces an error by a factor of order unity. With this assumption, the mean velocity downwards is half the terminal velocity, that is $g\lambda/2\bar{c}$. Further, let us put in the expression for n. This allows us to evaluate the nv_z term completely. Finally, by differentiating we evaluate the dn/dz term. We find

$$0 = -\frac{g\lambda}{2\bar{c}}n_0 e^{-mgz/kT} + D\frac{mg}{kT}n_0 e^{-mgz/kT}$$

which gives

$$D = \frac{\lambda kT}{2m\bar{c}}.\qquad(9.16)$$

If we put $kT = \frac{1}{2}m\bar{c}^2$ (not paying too much attention to the distinction between mean and r.m.s. speeds), we get

$$D = \frac{1}{4}\lambda\bar{c}.$$

The more exact expression arrived at in Appendix B is $\frac{1}{3}\lambda\bar{c}$. This is practically our result and the discrepancy is due to our crude averaging. This digression has therefore justified the idea of regarding the equilibrium of the gas as a balance between a drift velocity in one direction and a diffusion in the opposite direction down a concentration gradient.

Let us now return to the real subject of this section and use the same method for *particles suspended in a liquid*, undergoing Brownian motion—resin particles in water for example. We have already studied this system in section 4.4.2, when we concluded that the concentration is given by exactly the same law:

$$n = n_0 \exp(-m^*gz/kT) \qquad (4.17)$$

with m^* the effective mass, corrected for the buoyancy of the surrounding liquid.

Again in equilibrium we have a balance between drift downwards and diffusion upwards:

$$0 = -nv_z - D\frac{dn}{dz}. \qquad (9.17)$$

Let us assume here that the force on the particle is given by Stokes' law. Then the particle (of radius r) reaches a terminal velocity when the viscous force ($6\pi\eta rv_z$) is equal to the weight (m^*g): that is,

$$v_z = \frac{m^*g}{6\pi\eta r}.$$

Then our equation reads

$$0 = -\frac{m^*g}{6\pi\eta r}n_0 e^{-m^*gz/kT} + D\frac{m^*g}{kT}n_0 e^{-m^*gz/kT}$$

which gives

$$D = \frac{kT}{6\pi\eta r}. \qquad (9.18)$$

This is the diffusion coefficient of the particles of radius r through a medium of viscosity η at temperature T. Finally, let us assume that the same approach is valid when the particle is an atom of the liquid itself. As we have presented it, this is a rather controversial step to take. There are two points which we have to consider. First, are we justified in applying Stokes' law for the force on the moving atom? In deriving this law we clearly regarded the sphere as immersed in a continuum, a medium without any atomic structure, in which it generated stream lines. This hardly seems justified if the sphere itself is of atomic size. But the mean free time between collisions is certainly very short—in contrast to the situation in a gas—so that a continuum is not too bad an approximation. Further, we have seen that Stokes' law implies that the penetration depth of the disturbance caused by the passage of the sphere is between 1 and 2 atomic radii: this is likely to be roughly correct, so that perhaps Stokes' law *can* be applied. Secondly, are we justified in talking of the effective mass m^* of an atom inside a liquid of identical atoms? An *incompressible* liquid in a gravitational field has the same density at all heights (that is, the scale height is infinite) and this is equivalent to saying that m^* is zero. Admittedly, m^* does not appear in the final relation, but it is always dangerous to cancel both sides of an equation by a quantity equal to zero. Perhaps the best justification of this step is to say that real liquids are compressible so that there *is* a small change of density with height and m^*, though very small, is not zero.

We say then that the relation

$$D = \frac{kT}{6\pi\eta r} \tag{9.18}$$

where r is the radius of an atom, relates the diffusion coefficient and the viscosity of a liquid. This is called Einstein's relation between the two quantities.

★ ### 9.7.3 Viscosity of liquids

Using the expression for the variation of diffusion coefficient with temperature

$$D = D_0 \, e^{-A_0/kT} \tag{9.11}$$

where A_0 is the activation energy per atom, Einstein's relation gives

$$\eta = \frac{kT}{6\pi r D_0} e^{+A_0/kT}. \tag{9.19}$$

We can test this relation by comparing with experiment. It predicts that the viscosity *decreases* as the temperature is raised, in contrast to the behaviour of gases.

Among the most interesting measurements of the viscosity of liquid argon are those shown in Fig. 9.19. They were performed in a very simple way. The liquid was contained in a metal tube, through which a cylindrical weight could fall very slowly, the clearance between tube and cylinder being only small. The cylinder had a magnetic core inside it and its position could be detected magnetically. The time required to fall a known distance was determined. This was calibrated by timing the fall of the cylinder using liquids of known viscosity. The apparatus was robust and, as we shall see, was used up to very high pressures.

On the diagram, the two vertical dashed lines represent the triple point (83°K) and the critical temperature (150°K). The region between them represents the range of temperature in which the liquid can exist.

The full curve AB represents the viscosity of argon in equilibrium with its vapour pressure. At A, the liquid is at a low pressure, has a density of 1.37 g/cm^3, at B it is under 48 atmospheres and its density is 0.70 g/cm^3.

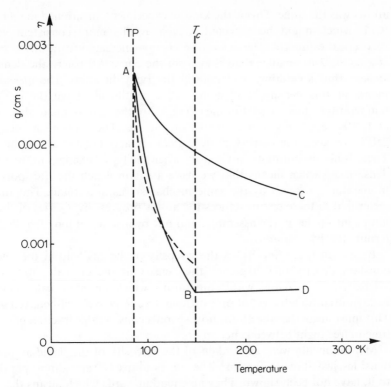

Fig. 9.19. Viscosity of liquid argon as a function of temperature. Line AB: measurements on the liquid under its saturated vapour pressure, from the triple point to the critical point. Dashed line: calculated from D using Einstein's relation, Eq. (9.18). Line AC: measurements at constant density, 1.37 g/cm³—the part to the right of the critical point refers to the gas. The descending curve shows liquid-like behaviour. Line BD; measurements on gas at constant density, $\rho = 0.70$ g/cm³. Data from Zhdanova, *Soviet Phys. J.E.T.P.* **4**, 749 (1957).

The dashed curve close to it is calculated using the experimental values of D (section 9.7) and Einstein's relation, assuming an atomic radius of 2 Å (or, what comes to the same thing, a rather smaller radius together with a larger number than 6π in the expression for the force on a moving atom). The agreement is good, and it shows that Einstein's relation is valid within a factor close to unity. The upper curve AC represents measurements at constant density equal to 1.37 g/cm³ (the density at the triple point temperature). For each measurement, the temperature was first raised to a chosen value and then the pressure was increased until the density reached the value 1.37 g/cm³; the viscosity was then determined. Thus for all points along this curve, the mean distance between atoms was the same. This is the kind of experiment, mentioned in section 9.6.1, which might be expected to give results more consistent with theoretical estimates. The activation energy calculated from this curve* is a factor 2 or smaller than that from the curve AB where the density varies—this is certainly a change in the right direction. The measurements at this density, 1.37 g/cm³ were continued above the critical temperature; thus the part of the curve AC to the right of the dashed line at 150°K represents measurements on the *gas*. The pressure needed at 250°K to keep the density at 1.37 g/cm³ is about 2,000 atmospheres. There is no discontinuity in viscosity when the liquid changes to the very dense gas—which means that we have a gas in which the transport of momentum proceeds by the same mechanism as in a liquid. This must mean that in these circumstances the atoms progress by a series of short jumps into holes in the assembly, and not by a series of long free flights terminated by collisions.

By contrast, the line BD is the viscosity of the gas held at the lower constant density of 0.70 g/cm³, the density at the critical point. The viscosity now *increases* with temperature, which we might call gas-like behaviour although η is not proportional to $\sqrt{T}$ as in a well-behaved gas. This must mean that the atoms now spend a considerable fraction of their time in free flight between encounters.

Measurements were also taken of the viscosity of the *liquid* at small densities, just above 0.70 g/cm³. The curves, above BD and almost parallel to it, have not been shown. They also tend upwards which means that in the liquid near the critical point there are sufficient large holes in the structure for the motion of the molecules to be gas-like.

PROBLEMS

9.1. (a) Estimate the extra energy per unit length (in erg/cm) of an edge dislocation in a metal due to the low coordination and increased interionic spacing of some of the ions.

(b) A specimen of metal has 10^9 dislocations per cm² of cross section. How much energy is stored in the dislocations per cm³? Will this extra energy appear as a reproducible contribution to the specific heat, like that due to vacancies—in other words, is the number of dislocations a unique function of the temperature? If not, why not?

(c) Work hardening is due to the presence of large numbers of dislocations tangled together and repelling one another, so that large stresses must be applied before they can move. The extra stress (dyn/cm²) must be of the same order as the energy density (erg/cm³) due to the dislocations. Estimate the concentration of dislocations at which significant work hardening takes place in a typical metal at a strain of 1%.

* Determined by plotting $\log(\eta/T)$ against $1/T$.

9.2. X-ray examination of a small region of a crystal of cubic symmetry and lattice parameter 4 Å shows that it consists of two slightly misaligned crystals. The angle between the planes is $\frac{1}{20}°$. What pattern of etch pits would be expected along the grain boundary? See Fig. 9.7(d).

9.3. The mobility q of a charged particle (such as a vacancy or an interstitial ion in an ionic solid) is defined by the equation

$$q = \frac{\text{mean drift velocity}}{\text{electric force } Ee \text{ acting on charge}}$$

where E is the electric field and e is the charge on the particle. The problem is to show that there is a relation between q and the diffusion coefficient D, using a method analogous to that in section 9.7.2.

(a) Consider a steady electric field E acting across a slab of material, considered to be a continuum without atomic structure. A positive charged particle inside the slab is attracted to the negative face and vice versa. What is its potential energy at a distance h from the face compared with the value at the face? In equilibrium, what is the probability of finding the particle between h and $(h+dh)$? Show that the charge is distributed exponentially with h, with a scale length equal to kT/Ee. Estimate this length at room temperature in a field of 1 volt/cm for e equal in magnitude to the electronic charge, remembering that kT at room temperature is $1/40$ eV.

(b) Does this result conflict with the predictions of elementary electrostatic theory about charge distributions?

(c) Set up a differential equation expressing the equilibrium as a balance between diffusion down the concentration gradient and a steady drift down the electric field. Apply this equation to the system discussed in (a). Deduce Einstein's relation: $q = D/kT$.

SOLUTIONS TO PROBLEMS

Chapter 3

3.1 (a) $-\dfrac{GMm}{r^2}$ (b) $-\dfrac{GMm}{r}$ (c) See Fig. P.1   (d) $g = \dfrac{GM}{a^2}$

3.2 (a) $\dfrac{2}{3}\dfrac{\mu_0 M^2}{h^3}$ (d) See Fig. P.2   (e) $\left(\dfrac{2\mu_0 M^2}{mg}\right)^{1/4}$

(f) $2\pi\sqrt{\dfrac{2^{5/4}}{8g^{5/4}}\left(\dfrac{\mu_0 M^2}{m}\right)^{1/4}} = \pi\left(\dfrac{2}{g^5}\dfrac{\mu_0 M^2}{m}\right)^{1/8}$

3.3 (d) See Fig. P.3.

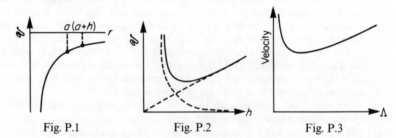

Fig. P.1   Fig. P.2   Fig. P.3

(e) $\dfrac{\text{surface tension energy}}{\text{gravitational energy}} = \dfrac{4\pi^2\gamma}{\Lambda^2\rho g}$, where $\gamma \sim 10$ dyn/cm, $\rho \sim 1.4$ g/cm^3,

$g \sim 10^3$ cm/s^2; 210 c/s; 21 cm/s.

3.4 28; 0.81 g/cm^3; 34.5 cm^3; ~ 3.9 Å; 8.8 dyn/cm; 0.0034 eV

3.5 (a) 28.5 cm^3, 3.55 Å   (b) 3.95 Å   (c) 0.0134 eV (note: $\mathfrak{n} = 12$)
     (d) 0.0114 eV   (e) 0.0124 eV
     (f) 2.5×10^{10} dyn/cm^2, compared with $\sim 2.2 \times 10^{10}$ erg/cm^3 or 1.7×10^{10} erg/cm^3 according to value of L_0.
     (g) 83 or 66   (h) $\sim 1.1 \times 10^{12}$ c/s

3.6 Assume close-packed discs in planes. $\mathfrak{n} = 6$, $\varepsilon = E/30$ *within* planes; $\mathfrak{n} = 2$, $\varepsilon = $ *between* planes.
     (a) $L_0 = 1.1NE$ per mol
     (b) If discs parallel to surface, $\mathcal{N} = (1/3\sqrt{3}r^2)$ (see Fig. P.4; each triangle contains $\frac{1}{3}$ disc and has area $\sqrt{3}r^2$) and $\gamma \sim 0.1E/r^2$. If perpendicular to surface, $\mathcal{N} = 1/(2r \times r/10)$, and $\gamma = 0.25 E/r^2$. Surface tension somewhere between these two.

Fig. P.4

(c) Tendency for parallel orientation because lower surface energy.

3.7 $a_0 = 2.32$ Å $N = 5.8 \times 10^{23}$ assuming r^{-10} repulsions.

3.8 Binding energy $= \dfrac{N\alpha e^2}{a_0}\left(1-\dfrac{\rho}{a_0}\right), \rho = 0.37\,\text{Å}$

3.9 (a) 4.55 eV

    (b) If solvents were structureless media, binding energy would be reduced by factor 80, 25, 18 respectively. But to screen one ion from another, solvent molecules must crowd round ions and size is important.

3.10 (a) 2.404   (b) 1.803

3.11 12 Å ; 9 Å ; 3.3 Å ; 7 Å

3.12 $K = 22.1 \times 10^4\,\text{kg/cm}^2$.

    Comparison with $\dfrac{P}{K} = \left\{\dfrac{3}{10}\left(\dfrac{V_0}{V}\right)^{14/3} - \left(\dfrac{V_0}{V}\right)^{4/3}\right\}$

    shown in Fig. P.5. Theory should be quite accurate.

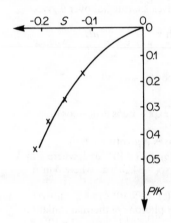

Fig. P.5

3.13 $\dfrac{1}{2\pi}[6a_0 Y/m]^{1/2}$ gives $3.4 \times 10^{13}, 0.86 \times 10^{13}$ and 1.56×10^{12} c/s.

Chapter 4

4.1 Effective mass 3.8×10^{-14} g, scale height 10.0 μ. $N = 6.47 \times 10^{23}$.
    Inverted concentration gradient with same scale height.

Chapter 5

5.1 (a) 0.0022 Å

    (b) 0.0104 Å if other causes of broadening are independent of temperature.

5.2 (a) $-\dfrac{GMm}{r^2}; -\dfrac{GMm}{r}$ (b) $\exp\left(+\dfrac{GMm}{rkT}\right)$

(c) $A4\pi r^2 \exp(GMm/rkT)\,dr$, where A is a normalizing factor.

(d) exp factor $\to 1$, so that whole expression $\to \infty$ when $r \to \infty$.

(e) (ii); (c) cannot be normalized.

5.3 Centre of gravity rises by R/Mg for 1° increase of temperature so that gravitational potential energy increases by R.

5.4 31% assuming that at 1 atm, both N_2O_4 and NO_2 act like perfect gases.

5.5 11.2 km/sec; 1.37 km/sec.

High-speed 'tail' of Maxwell distribution allows more He atoms than O_2 or N_2 atoms to escape; see problem 2 above, part (e).

5.6 (a) $(u_x - 2\xi)$; $\frac{1}{2}mu_x^2 - 2mu_x\xi$

(b) $nAu_x\,d\tau \cdot P[u_x]\,du_x$; $nAu_x\,d\tau \cdot P[u_x]\,du_x(-2mu_x\xi)$;
$-2nm\,dV \int_0^\infty u_x^2 P[u_x]\,du_x = nkT\,dV$

(c) $\frac{3}{2}knV$; $dT = -dVkT/\frac{3}{2}knV$

(d) $TV^{2/3} = $ const., becomes $PV^{5/3} = $ const. since $PV = RT$

(e) $dT = -d\,VkT/C_vV$ where C_v is specific heat per molecule.

(f) See discussion in section 5.1.2.

5.7 (i) $A \sin\theta\,d\theta\,d\phi \exp(-(\mu H/kT)\cos\theta)$, A a normalizing factor.

(ii) Integral over ϕ gives 2π. Integral over θ gives

$$\frac{2\sinh x}{x} \quad \text{where} \quad x = \frac{\mu H}{kT}, \quad \text{whence} \quad A = \frac{x}{4\pi} \cdot \frac{1}{\sinh x}.$$

(iii) $\dfrac{x}{2\sinh x} \exp(-x\cos\theta) \cdot \sin\theta\,d\theta$

(iv) $N\mu \cos\theta \dfrac{x}{2\sinh x} \exp(-x\cos\theta)\,d(-\cos\theta)$

(v) Integral over θ gives $N\mu[\coth x - 1/x]$

5.8 (a) 5×10^{12} c/s.   (b) $1.89 \times 10^{-3}\delta^2$ J, where δ in Å

(c) $p[\delta]\,d\delta = 0.304 \exp(-34.2\delta^2)\,d\delta$ where δ in Å

(d) 0.12 Å

5.9 (a) 2   (b) kT (c) 2.9×10^{-4} rad   (d) (v)   (e) (iii)

(f) Bombardment by photons of thermal radiation.

5.10 (a) 6×10^{-39} g cm^2 (b) $\frac{1}{2}kT$ (c) 3×10^{12} c/s

(d) 6 vibrational + 3 rotational degrees of freedom; $\frac{9}{2}R$

(e) 3 translational + 3 rotational degrees of freedom; $\frac{6}{2}R$

(f) $\frac{8}{6}$

(g) Rotations (c) take place, rotation about dumbell axis does not; 5 degrees of freedom; γ near $\frac{7}{5}$.

5.11 1.4

5.12 0.3 eV   (b)

Chapter 6

6.1 (a) $C_v = 20.8$ J/mol deg, $\sigma \sim 35 \times 10^{-16}$ cm^2, $\bar{c} \sim 4.5 \times 10^4$ cm/sec; $\kappa \sim 1 \times 10^{-4}$ watt/cm deg

(b) 0.175 watt   (c) 2×10^{-3} m

(d) $\kappa = \frac{1}{3}n(C_v/N)\bar{c}d$ where d is spacing; 2×10^{-4} m

6.2 (a) $-(DA/h)mn_v$

(b) $-\rho_L A \, dh/dt$ g/s where A is area of tube and ρ_L is density of liquid

(c) $h^2 = \dfrac{2Dn_v m}{\rho_L} t$

(d) 0.11 cm^2/s. Assume collisions with air molecules are much more frequent than with other ether molecules; then $\lambda = 1/n_A \sigma_E$, where subscript A means air, E ether; $D_{EA} = \frac{1}{3}\lambda \bar{c}_E$; whence $\sigma_E = 10^{-15}$ cm^2.

6.3 Assume only collisions with air molecules are important; $\lambda = 1/n_A \sigma$ where n_A = number of air molecules/cm^3. Take $\sigma = 10 \times 10^{-16}$ cm^2. For small thermionic tube, dimension $\sim$ 1 cm, volume $\sim$ 1 cm^3; for large transmitting tube, dimension $\sim$ 10 cm. volume $\sim 10^3$ cm^3. Then $P \sim 0.02$ mm, 0.002 mm and area of hole $\sim 10^{-11}$ cm^2, 10^{-12} cm^2 respectively (Knudsen flow).

6.4 Maxwell distribution inside vessel. All molecules within cone of height c and solid angle determined by shape and size of hole will leave vessel in 1 sec; whence factors c and G respectively.

$$\text{Mean kinetic energy} = \frac{G \int_0^\infty \frac{1}{2}mc^2 c P[c] \, dc}{G \int_0^\infty c P[c] \, dc}.$$

Denominator is equivalent to $\frac{1}{4}n\bar{c}$.

Expression $GcP[c]\,dc$ for number escaping per sec shows that fast molecules predominate and distribution is non-Maxwellian.

Chapter 7

7.1 (a) $T_i = \frac{27}{4}T_c$; 220°K

(c) $T_B = \frac{1}{2}T_i = 110°$K, so that at room temperature $PV > RT$.

7.2 $V_c = 2b$; $T_c = \dfrac{a}{4Rb}$; $P_c = \dfrac{a}{4e^2 b^2} \sim \dfrac{a}{30b^2}$; $B(T) = \left(b - \dfrac{a}{RT}\right)$; $T_B = \dfrac{a}{Rb}$

7.3 (a) 200 cm^3; 4.4 Å

(b) B plotted against $1/T$; $B = 280 - (2.18 \times 10^5)/T$ is a rough fit—see Fig. P.6.; $a \sim 1.83 \times 10^6$ J cm^3/mol; using $a/b \sim 2.7N\varepsilon$, $\varepsilon \sim 0.025$ eV.

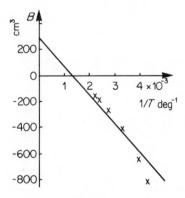

Fig. P.6

(c) 0.036 eV

(d) $\mathcal{V} = \left(\dfrac{\lambda}{r^{12}} - \dfrac{\mu^2}{4\pi\varepsilon_0 r^3} \right)$ gives $\mathcal{V}_{\min} = -\varepsilon = -\dfrac{3}{4} \cdot \dfrac{\mu^2}{4\pi\varepsilon_0 a_0^3}$; 0.8×10^{-29} C m.

(e) 1.6×10^{-20} C m; $\pm\frac{1}{4}$ electron charge 2 Å apart, or larger charges nearer together.

7.5 (a) zero   (b) r.m.s. distance $= \sqrt{Nl}$ (c) $D = \frac{1}{2}(l^2/\tau)$

7.6 (a) $\bar{x} = \dfrac{Mg}{\alpha}$ (b) $\sqrt{\dfrac{kT}{\alpha}}$ (c) $\dfrac{\sqrt{(kT\alpha)}}{g}$

Chapter 8

8.1 6; 4.82 Å; 16.1°, 33.6°, 56.1°

8.2 Bragg's law gives $dd/d = -d\theta/\tan\theta$ for small changes of d and θ; $\alpha = 10^{-5}$. Expansion coefficient varies with direction.

8.3 (b) 16×10^{-6}, for $p = 11$

(d) $\gamma = 0.85$ at 30°K, 1.28 at 65°K, 1.45 at 283°K compared with 1.67 for $p = 11$.

Chapter 9

9.1 (a) $\sim 10^{-4}$ erg/cm

(b) 10^5 erg/cm^3. No, number depends on mechanical and thermal treatment of specimen.

(c) Elastic energy $= (K/2)s^2 \sim 10^8$ erg/cm^3; 10^{12} dislocations/cm^2

9.2 DD^1 (Fig. 9.7(d)) is given by $a/\text{DD}^1 = \sin\theta$ where θ is angle between grains, a is lattice parameter. Line of etch pits 4460 Å apart.

9.3 (a) Veh/l; $Ve/lkT \exp(-Veh/lkT)\,dh$, if $l \gg h$; $\frac{1}{40}$ cm.

(b) Yes. Elementary electrostatics neglects the diffusion of electrons, because electric charge is treated as a continuous fluid.

(c) $D\,dn/dh + qn(Ve/l) = 0$, where $n = $ no. of charged particles/cm^3; this gives $n \propto \exp[-q(Veh/lD)]$.

Reading List

The following list contains a few suggestions for complementary and further reading.

GENERAL

R. P. Feynman, R. B. Leighton and M. Sands, *The Feynman Lectures on Physics*, Vols. 1 and 2, Addison-Wesley, Reading, Mass., 1963.

T. L. Hill, *Lectures on Matter and Equilibrium*, Benjamin, New York, 1966.

F. Mandl, *Statistical Physics* (Manchester Physics Series), to be published by John Wiley, London.

F. Reif, *Statistical Physics* (Berkeley Physics Course, Vol. 5), McGraw-Hill, New York, 1967. (Mainly relevant to chapters 4–9 of this book.)

F. O. Rice and E. Teller, *The Structure of Matter*, Science Editions, New York, 1966.

D. Tabor, *Gases, Liquids and Solids*, Penguin Books, Harmondsworth, Middlesex, 1969.

M. W. Zemansky, *Heat and Thermodynamics*, 5th ed., McGraw-Hill, 1968.

CHAPTER 3

A. H. Cottrell, *The Mechanical Properties of Matter*, Wiley, New York, 1964.

C. Kittel, *Introduction to Solid State Physics*, 3rd ed., Wiley, New York, 1966.

CHAPTER 4

R. D. Present, *Kinetic Theory of Gases*, McGraw-Hill, New York, 1958.

E. Schrödinger, *Statistical Thermodynamics*, Cambridge University Press, Cambridge, 1964.

H. D. Young, *Statistical Treatment of Experimental Data*, McGraw-Hill, New York, 1962.

CHAPTERS 5, 6 and 7

R. D. Present, *Kinetic Theory of Gases*, McGraw-Hill, New York, 1958.

CHAPTER 8

B. E. Chalmers, *Principles of Solidification*, Wiley, New York, 1964.
C. Kittel, *Introduction to Solid State Physics*, 3rd ed., Wiley, New York, 1966.
F. C. Phillips, *An Introduction to Crystallography*, 3rd ed., Longmans, London, 1966.

CHAPTER 9

F. P. Bowden and D. Tabor, *Friction and Lubrication*, Methuen, London, 1967.
B. E. Chalmers, *Principles of Solidification*, Wiley, New York, 1964.
A. H. Cottrell, *The Mechanical Properties of Matter*, Wiley, New York, 1964.
J. Frenkel, *Kinetic Theory of Liquids*, Dover, New York, 1955.
C. Kittel, *Introduction to Solid State Physics*, 3rd ed., Wiley, New York, 1966.

The following films deal with experiments which are described in this book. During the course of each film, *measurements* can be taken from the screen so that students can write down their own readings and later work out their own results. A booklet is available giving essential numerical and other data about each piece of apparatus.

The series is called 'EXPERIMENT'.

No. 5	The Determination of Boltzmann's Constant	(Chapter 4)
No. 3	C_p/C_v for Helium, Nitrogen and Carbon Dioxide	(Chapter 5)
No. 4	The Effect of Pressure on the Thermal Conductivity of a Gas	(Chapter 6)
Nos. 1 and 2	pV Isotherms of Carbon Dioxide	(Chapter 7)

Films (16 mm sound colour, 15 minutes each) can be purchased from Granada International Productions Ltd., 36 Golden Square, London W1, or hired from The British Film Institute Distribution Library, 42/43 Lower Marsh, London SE1.

Index

References such as 304(9.1) are to Problems, where information is given which is not in the text.